守护城市密度
Defending Density

[荷]大都会建筑事务所等 | 编

于风军 林英玉 蒋立真 曲洪波 张可新 辛敏裕 侯逸宁 | 译

大连理工大学出版社

004 旅游、建筑、新常态 _ Fabrizio Leoni

新项目

008 Schedlberg沉思之家 _ Peter Haimerl . Architektur

026 法布拉和科茨老工业建筑 _ Roldán + Berengué, arquitectes

044 布莱顿学院体育与科学系大楼 _ OMA/Ellen van Loon

保持社交距离，守护城市密度

058 **保持社交距离，守护城市密度** _ Richard Ingersoll

066 Carlo Erba住宅楼 _ Eisenman Architects + Degli Esposti Architetti + AZstudio

082 Fenix一期项目 _ Mei architects and planners

098 Bottière Chénaie住宅综合体 _ KAAN Architecten

118 Denizen Bushwick公寓楼 _ ODA

幼儿园：灵活、安全

132 **幼儿园：灵活、安全** _ Paula Melâneo

138 Würenlingen幼儿园 _ Malte Kloes Architekten + Estrada Reichen Architekten

150 Storey's Field中心与爱丁顿托儿所 _ MUMA

164 06A地块幼儿园 _ sam Architecture + Querkraft

178 MUKU托儿所 _ Tezuka Architects

"如水一样，我的朋友"

190 **"如水一样，我的朋友"** _ Diego Terna

198 博登公园自然游泳池 _ gh3* architecture

210 水上公园 _ Ferrier Marchetti Studio

222 Termalija家庭健康中心 _ ENOTA

236 建筑师索引

004 Tourism, Architecture and New Normality _ Fabrizio Leoni

New Projects

008 Schedlberg, Contemplation House _ Peter Haimerl.Architektur

026 Fabra & Coats _ Roldán + Berengué, arquitectes

044 Brighton College Sports and Science Building _ OMA/Ellen van Loon

Defending Density in the Year of Social Distancing

058 **Defending Density in the Year of Social Distancing _ Richard Ingersoll**

066 Carlo Erba Residence _ Eisenman Architects + Degli Esposti Architetti + AZstudio

082 Fenix I _ Mei architects and planners

098 Bottière Chénaie Housing Complex _ KAAN Architecten

118 Denizen Bushwick _ ODA

Kindergarten: Being Flexible, Being Safe

132 **Kindergarten: Being Flexible, Being Safe _ Paula Melâneo**

138 Kindergarten Würenlingen _ Malte Kloes Architekten + Estrada Reichen Architekten

150 Storey's Field Center and Eddington Nursery _ MUMA

164 O6A Lot Kindergarten _ sam Architecture + Querkraft

178 MUKU Nursery _ Tezuka Architects

"Be Water, my friend"

190 **"Be Water, my friend" _ Diego Terna**

198 Borden Park Natural Swimming Pool _ gh3* architecture

210 Water Park Aqualagon _ Ferrier Marchetti Studio

222 Termalija Family Wellness _ ENOTA

236 Index

旅游、建筑、新常态
Tourism, Architecture and New Normality

Fabrizio Leoni

在旅游的通用标签下，我们似乎可以以主办或组织与招待、游牧生活、临时主题住房有关的多种活动为目的设计空间。尽管旅游业起源很早，但是旅行者的形象是从现代才开始呈现的。他们是文化、政治领域里的精英，在前一代全球化的环境中，因为精通拉丁语、法语和后来的英语，他们可以在地理中心之间顺畅地活动，并且给当地带来广泛的知识。从中世纪末开始，为了巩固殖民帝国，音乐家、诗人、演员、作家、外交家、建筑师、军事战术专家、政治顾问就前往欧洲、非洲和亚洲，开始了与商业往来有关的职业游牧活动。宗教领头人和传教士不断在各地现身，长期致力于传播思想和精神价值，并通过持续地传教和朝圣体现对新大陆的"文化征服"。生产时间和冥想时间要有明确的界限，这是他们旅行的第一个目标。18和19世纪的大巡演把直接接触意大利丰富艺术遗产与休闲活动联系起来，将知识与愉悦融合在一起。20世纪30年代，国际建筑现代化协会会员乘坐"帕特里斯II号"轮船在地中海巡游，从马赛开往雅典，将这样的一种融合变成了一件普通的事。

同时，健康水疗旅游引入了康复的范式，它是上层阶级生产生活的"临时休假"。这些人有足够特权将保持身心健康作为一种价值观。山区、海滨和城市中心的大酒店是这些活动的具体结果，还有温泉浴、海水保健诊所以及后来随着大众开始关注医疗保健而在世界各地修建的疗养院或度假中心：从帕米奥疗养院到

Under the generic label of Tourism, it seems we can include the set of spaces oriented to host and organize the multiple activities related to hospitality, nomadic living, thematic temporary housing. Although tourism has very remote origins, it is in a modern perspective that the figure of the traveller emerges. It is somebody belonging to the cultural, political elites, who in an environment of ante litteram globalization and favoured by the standard mastering of Latin, French and, later on, English, smoothly moves between the geographical centers, bringing with him a wide range of knowledge. Musicians, poets, actors, writers, diplomats, architects, experts in military tactics, political advisers, travel since the end of the Middle Age to the consolidation of colonial empires, to European, African and Asian territories, unfolding professional nomadism linked to commercial traffic. The mobile presence of religious leaders and missionaries has been long-term committed to spreading ideas and spiritual values, and to embodying the "cultural conquest" on new lands over a continuous action of proselytism and pilgrimage. The boundary between productive and contemplative time is clear, being the first and the main object of the journey. The Grand Tour, at the eighteenth and nineteenth century, introduces a mix of intellectual commitment and pleasure, associating the direct contact with the vast Italian artistic heritage to the mere loisir. The members of Congrès Internationaux d'Architecture Moderne cruising the Mediterranean sea on the steamer Patris II, sailing from Marseille to Athens, in the 1930s, turn into ordinary such a merge.

In parallel, the health and spa tourism introduces the paradigm of healing as a "temporary leave" from productive life for the upper class, those privileged enough to frame their psychophysical well-being as a value to be preserved. The Grand Hotels in the mountain areas, at the seaside and urban centers, are the tangible results of these events, along with thermal baths, thalassotherapy clinics and those structures that, later, with the ongoing process of extension of healthcare to the general public start to open all over: from the Paimio Sanatorium to the marine colonies during Fascism in Italy, from the

意大利法西斯主义时期的海洋殖民地，从新以色列的疗养院到为苏联工人的娱乐活动而建立的度假中心。

一个世纪以来，一个故事的发展推动着旅游概念发展成为一个系统，这个故事起源于以前为全球的富人提供假期的独特概念，这种独特概念借用自文学作品（斯科特·菲茨杰拉德的《夜色温柔》）、影视作品（《罗马假日》《威尼斯之死》）和以杂志（《Vogue》）和电视为媒介的当代大众文化。旅游系统中所有可能的组成元素（从商业活动到纯粹的休闲活动，从医疗保健到文化利益）都日趋成为一种"整合的服务"，一种综合了承诺与休息、紧凑时间与松散时间的系统。

此时，类型化的供应链就这样形成了，它将偶尔的或季节性的活动以及所有可能的功能组合从手工活动转变为工业活动，推动了建筑和景观设计、通信、家具和时装设计、基础设施工程、广告等的聚合，为游牧民族提供了稳定的聚居地。从博尔卡迪卡多雷的埃尼村（服务范围广泛的企业旅游的代表）精制有机理性主义到意大利翡翠海岸的喷气式飞机旅游的语言模仿，从著名的法国地中海度假俱乐部到高山滑雪和水上运动的专业化综合建筑，所有这些都使当地发生了巨大的实质性变化。

大众传媒孕育了特定的图像学，产生了神话、图像、故事以及文学和视觉艺术。今天无处不在的社交媒体通过影响和更新已经在城市规划和建筑手册中沉淀下来的物理范式，丰富和空间化了游牧生活方式的概念。

Houses of Convalescences in the new State of Israel to the holiday centers promoted for the recreation of the Soviet worker.

Over the century, the evolution of a narrative originating from a previously exclusive idea of a vacation for the affluent and the cosmopolitan, borrowed through literature (Scott Fitzgerald's Tender is the Night), cinema (Grand Hôtel, Roman Holiday, Death at Venice) and an incumbent mass culture mediated by magazines (Vogue) and TV boosts a concept of Tourism morphed into a systemic dimension, in which all its possible components, from business to pure leisure, from healthcare to cultural interests, tend to become an "integrated offer", a hybrid of commitment and rest, dense and rarefied time.

At this point, the typological supply chain completes, transforming occasional or seasonal activities from artisanal to industrial, with all possible mix of programs, pushing architecture and landscape design, communication, furniture and fashion design, infrastructure engineering, advertising, to converge and generate stable habitats for the nomadic citizen. From the refined organic rationalism of the Eni Village in Borca di Cadore, an icon of inclusive corporate tourism, to the linguistic pastiche of the jet-set tourism of Costa Smeralda, Italy, from the acclaimed French resorts Club Med to the specialized complexes for alpine skiing and water sports, all of them drive significant physical changes to the places.

Mass media gave birth and fed a specific iconography, producing myths, images, storytelling, together with literature and visual arts. Today's pervasive social media amplify and spatialize the concept of a nomadic lifestyle fuelled by the programs as mentioned above, by influencing and updating the physical paradigms already sedimented in urban planning and architectural manuals.

As the epidemics or its perceived danger abruptly spreads over the whole planet, such a nomadic living that we were used to identifying with those lifestyles is likely to change. What if the emergency

随着流行病及其所带来的危险突然蔓延到整个地球，我们过去所认同的那些游牧生活方式很可能会发生改变。如果应急措施克服了临时性，变成一种持久的范式或"新常态"，会怎样呢？旅游限制条例、社交距离、空间关系规则和安全范式所提出的新空间规范要求会如何触发不同的社会互动和不同的旅游形式？它们会影响固有的思维定式、减少活动空间或者损害个人权利吗？

我们不应忽视这一点——旅游业和其他行业活动一样，提供综合性的活动，如今被归类为颠覆性产业。旅游是利用一个地区的自然、历史、文化等"原始资本"组织的活动，既可以是积极的也可以是消极的，其颠覆性就在于它的两极分化——既有消极的意义，也有积极的意义。一方面，旅游业会"制造麻烦"，从而使当地原有的模式无法延续"；另一方面，旅游业会"改变一个行业的传统运作方式，使其变得更加新颖和高效"。旅游业有强大的多层次的创新驱动力，无论这种短暂的或永久性的活动发生在哪里，其巨大的空间需求都会立即显现出来。因此，旅游业主要模式快速和相对不可预知的变化会使整个服务文化变得非常不稳定，这也是旅游业持续反复变化的原因，包括从富裕阶层专享的服务/产品（低密度）到社会普遍享有的系统性服务（高密度）的改变，这种现象需要对专用空间进行频繁升级，从而用较高级的手段废弃这些专用空间。所有这一切听起来都非常相似，而且与流行病导致的一系列政策突变兼容：旅游业本来就是一个既能平稳发展又能迅速变化的行业，我们已经完全习惯了这种灵活性。

如果旅游，如前所述，不仅仅是任何人日常生活中的一个非生产性的活动，也不仅仅是由纯粹的休息的欲望触发的，而是一种主题性的、巡回式的生活方式，那么，各种各样的服务项目如何适应新的条件呢？丰

measures overcome temporality and turn into a long-lasting paradigm or "new normality"? How the new normative spatial requirements, issued by travel restrictions, social distancing, proxemic rules, safety paradigm, would trigger different social interactions and different forms of tourism? Would they threaten acquired mindsets and reduce spaces for action or jeopardize personal rights?
We should not ignore that as other industrial activities, Tourism, with its integrated offer of a multi-dimensional program, is classifiable today as a disruptive industry. It is an activity organized to take advantage of an "initial capital" – natural, historical, cultural – available in a territory, both negatively and positively, disruptive in its polarized negative/positive double meaning of "causing troubles and therefore stopping something from continuing as usual" and "changing the traditional way that an industry operates, especially in a new and effective way". The industry as a robust multi-level driver of innovation, whose insatiable spatial demand is immediately visible wherever such transient or permanent activities take place. That's why a rapid and relatively unpredictable change of its main patterns makes the whole culture of hospitality very volatile and responsible for a consistent shift back and forth, ranging from a service/product targeting exclusivity for affluent classes (low density) to a systemic dimension of a socially inclusive offer (high density): a phenomenon calling for frequent updates and a consequent high grade of spatial obsolescence of its dedicated spaces. All this sounds something very similar and compatible with the set of policy sudden changes due to the pandemic: tourism is inherently an industry subject to both smooth and rapid change, and already perfectly accustomed to adaptability.
If tourism, as said, is not merely an unproductive pause in any person routine, and is not purely ignited by a generic desire for rest, but a thematic, itinerant form of living, how the wide array of hospitality programs are subject to adaptation to new constraints? The rich typological catalogue of hospitality programs would suffer a drastic downsize? What about the named thematics: healing, sport, education, cultural interests or the many other reasons that urge us away from home?

富的服务项目是否会大幅减少？被命名的主题活动——疗伤、运动、教育、文化兴趣或其他很多促使我们离开家的原因——又会发生什么变化？

　　数字网络的扩大，尤其是通过预期的大规模公共和私人投资改善网络之后，人们有机会接触文化活动（运动、音乐会）、在大量博物馆馆藏中进行虚拟旅游、使用各个领域的学习教程、积极有效地利用业余时间。新形式的远程医学或疾病检测和控制，会促进物联网、地联网和其他数字化的进程。

　　一个关键问题是：建筑师有多大的兴趣参与进来，并且在设计中加入数字化元素？从另一个层面来讲，我们应该只在界定的地域范围内旅游，吸引游客在当地旅游，即所谓的近距离旅游。这样游客就会有兴趣了解当地或周边地区的知识。

　　我们要做的是重新平衡地区密度、恢复远离城市中心的中等规模城镇和村庄的发展机会。这是一个改变他们生活节奏和吸引游客的宝贵机会，不但暂住者（作为局外人）和当地居民（作为内部人）都会受益，还可以使当地资源持续发展，平衡各地区发展机会，提高生活质量。

　　今天，旅游业越来越关注人类福祉和文化。在这种新的背景下，针对城市贫困地区和被忽视的农村地区所采取的措施将在重塑自然环境方面发挥关键性的作用。要实施这些积极的措施，就要在建筑和基础设施的设计方面投资，而这反过来可能会影响我们城市的公共空间。城市需要改变专门用于汽车和私人交通的区域，使之成为行人、骑行人、社区活动和新的交通方式共享的混合区域。

The expansion of digital networks, especially if improved by expected massive public and private investments, would allow access to cultural events (sports, concerts), virtual tours of immense museum deposits, use of tutorials for learning out of the most varied fields, exploring an appealing alternative use of spare time. New forms of remote medicine or disease detection and control would emerge, likely promoting the Internet of Things, Internet of Places and any other digital fascinating advancement.

A crucial question could be: to which extent are architects interested in being engaged and will give a spatial response to these electronic moves? On a different level of the discourse, we should start to travel only within defined territorial units, attracted or subtly obliged to local tourism, the so-called tourism of proximity. A tendency to discover internal, neighboring worlds and regional knowledge would unfold.

Rebalancing territorial density, restoring opportunities to medium-sized towns and villages, which have suffered a sort of isolation from the main centers are what we need to do. It is a precious opportunity to recommend a change of speed, pace, and attention, useful not only for those who live in the area temporarily as outsiders, but also for those who live their everyday life as insiders, in a logic of sustainable enhancement of local resources and balanced dissemination of development opportunities and quality of life.

Today, tourism is increasingly shifting towards well-being and culture. In this new context, direct action on deprived part of cities and neglected rural areas will play a crucial role in reshaping the natural environments. Such a proactive approach would require a steady investment in design of both buildings and infrastructure which, in turn, would probably affect the public space of our cities. Cities will need to change where areas exclusively dedicated to cars and private transport will become hybrid areas enjoyed by pedestrians, bikers, community-based activities and new modes of transportation.

Schedlberg沉思之家
Schedlberg, Contemplation House

Peter Haimerl. Architektur

翻新破旧的巴伐利亚农舍，建成沉思空间，续写农舍的故事
A dilapidated Bavarian farmhouse is refurbished to continue its narrative as a contemplative space

巴伐利亚森林Schedlberg附近具有鲜明特点的传统农舍嵌在环境艰苦的东巴伐利亚景观中。现存的农舍已经很少了。人们想忘掉这个古老的、不富裕时代的象征，所以大部分农舍被拆除了，只有少数农舍变成了露天博物馆，还有一些被遗忘了，衰败了。

有一座曾经被当地的老农民使用的农舍，1963年被废弃。这座木屋，连同它的花岗岩基座，成了一片废墟。木屋的主人建立了新的家园，遗弃了原来的农舍。在附近吃草的牛羊把废弃的农舍当作了庇护所；真菌和蕨类植物在这里四处蔓延。只有起居区基本保持了原样，外墙仍然竖立着，支撑着屋脊檩条。房子看上去就像马上要倒塌了，这座介于自然与文化之间的建筑，显得弱不禁风。在这里，土壤和森林多于建筑本身，木屋就要消失了。

我们要在挖掘机将其铲平之前续写古老农舍的故事。农舍及其所在场地应该加入新的材料，与旧元素形成对比，而不是采用翻新的方式把废墟的秘密掩盖掉。农舍及其所在场地持续联结在一起，并没有隐藏农舍的房龄。大门外面长满苔藓的花岗岩用混凝土钢筋替代，建筑内有需要的地方也用混凝土支撑风化腐烂的木头。建筑的东面和南面，混凝土与木材相得益彰；建筑的背面，混凝土看起来像是悬浮的；建筑的东面，混凝土相互穿插，自由堆叠，重新排列。混凝土和平台互锁相连，将建筑变成了雕塑，将房子变成了艺术品。新老窗户共同框住了外面的景色。三居室的房子里提供了生活的必需品：烤箱、炉子、桌子、床和Wi-Fi。

农舍的主人和附近的农场主不同意出售农舍的废墟，他们不希望陌生人在此处逗留太久。高支出、低收益使得上述计划变得毫无价值；一旦建筑不再是投资的对象，也就失去了价值。渐渐地，房子成了农民的财产，投资者被赶走了。农舍的价值不会随着时间的推移而增加，反而会减少。我们要看的是农舍的现状，而不是以后的价值。现在，Schedlberg的这座建筑是能够让人冥想的建筑，游客可以在此参加研讨会、休养、反思。来访者不会破坏房子——他们会修缮房子。事实上，每一位来访的客人，都会使房子变得更有价值。思考的人暂住在房子里，会维护房子，确保它的价值。在30年的租赁期内，房子将作为一件艺术品得到维护资金。

The traditional farmhouses near Schedlberg in the Bavarian Forest, with their strong character, are embedded in the harsh, east Bavarian landscape. Only a few still exist today. As symbols of an older, less wealthy era, that people wanted to forget about, most of them were demolished, but a small number survived as open-air-museums. Others still were forgotten about and left to decay on their own.

One such house, previously used by the senior farmers in the area, was abandoned in 1963. The log house, with its granite basement, survived as a ruin. The inhabitants built a new home and left it in a state of dereliction. Cows and sheep, grazing in the adjacent meadow, used it as shelter; fungi and ferns sprawled. The living area alone remained largely untouched, the outside wall still erect, holding the ridge purlin. The house was on the verge of collapse; it was light, fragile, hovering between nature and culture. There are more soil and

项目名称：Schedlberg Contemplation House
地点：Arnbruck, Germany
建筑事务所：Peter Haimerl . Architektur
工作人员：Peter Haimerl, Jutta Görlich, Ulrich Pape, Tomohide Ichikawa, Maximilian Hartinger, Verena Höß Anne Zollner
建造方：HAUS.PATEN Bayerwald KG
结构工程公司：aka ingenieure
暖通空调：Consult ddk GmbH
施工：Spannagl Bau GmbH
建筑设备：Zelzer GmbH
可居住空间面积：approx. 180m²
造价：approx. €400,000
竣工时间：2018
摄影师：©beierle.goerlich (courtesy of the architect)

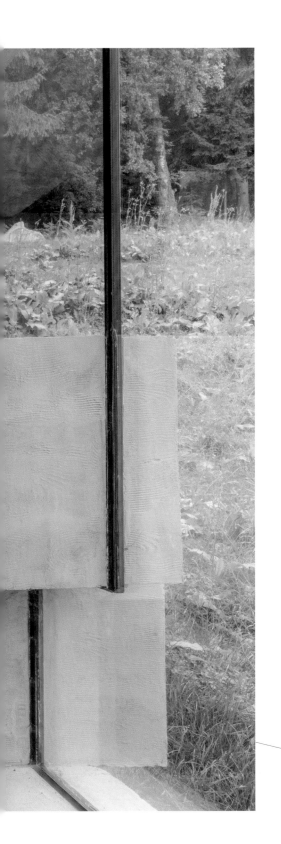

二层 first floor

一层 ground floor

1. 入口大厅 2. 起居室 3. 餐厅 4. 厨房 5. 原谷仓 6. 浴室 7. 储藏室/设备间
8. 大厅 9. 卧室 10. 更衣室 11. 冥想空间 12. 睡眠房间 13. 上空空间
1. entrance hall 2. living room 3. dining area 4. kitchen 5. former barn 6. bathroom 7. storage/services
8. hall 9. bedroom 10. dressing room 11. thinking space 12. sleeping room 13. void

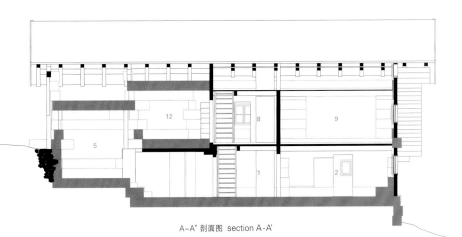

A-A' 剖面图 section A-A'

forest than architecture. It was about to disappear. There is a narrative that needs to be continued, before the digger is brought to bear. House and place are continued with new materials and contrasts, without refurbishing the secret of decay away. House and place are continuously entwined, the age of the house embraced, not hidden. Mossy granite bars, lying right outside the front door, were converted into concrete bars and implemented into the building where needed, supporting the weathered, decayed wood. In the east and south they complement the wood; on the north side they seem to fly; and on the east side they crash into each other, being freed, piled up and rearranged. The interlocking of bars and decks transforms the structure into sculpture, the house into art. Windows – old and new – frame the outside world. The three-bedroom house offers the true essentials: oven, stove, table, bed and Wi-Fi.

The owner of the property and the adjacent farm didn't agree to sell the ruin, not wanting strangers to stay too long. The high expenditure and low opportunities for profit made the scheme worthless to many; for them, once architecture is no longer an object for investment, it loses its meaning. Year by year and day by day the house settles even more into the possession of the farmer, driving investors away. Its value doesn't increase with time; it supposedly diminishes.

It is essential to see the building for what it is now, rather than what it is worth later. Now, the Schedlberg is contemplative architecture. It invites visitors in for seminars, retreat, reflection. Visitors don't wear the house down with their presence – they recharge it. In fact the house gets more valuable with each visitor. Thinkers temporarily live in the house, securing it and its value. In 30 years of endowed lease, the house will fund its own preservation as a piece of art.

法布拉和科茨老工业建筑
Fabra & Coats

Roldán + Berengué, arquitectes

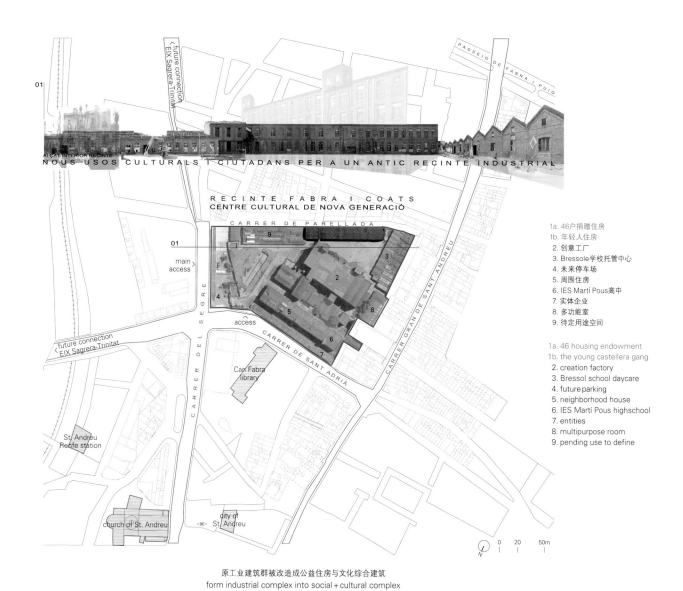

1a. 46户捐赠住房
1b. 年轻人住房
2. 创意工厂
3. Bressole学校托管中心
4. 未来停车场
5. 周围住房
6. IES Martí Pous高中
7. 实体企业
8. 多功能室
9. 待定用途空间

1a. 46 housing endowment
1b. the young castellera gang
2. creation factory
3. Bressol school daycare
4. future parking
5. neighborhood house
6. IES Martí Pous highschool
7. entities
8. multipurpose room
9. pending use to define

原工业建筑群被改造成公益住房与文化综合建筑
form industrial complex into social + cultural complex

法布拉和科茨丝线制造厂，圣安德鲁，1932年
The factory of thread manufacturer 'Fabra & Coats', Sant Andreu, 1932

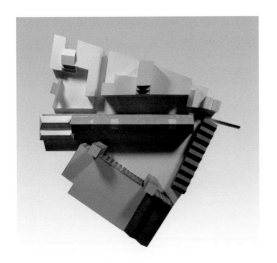

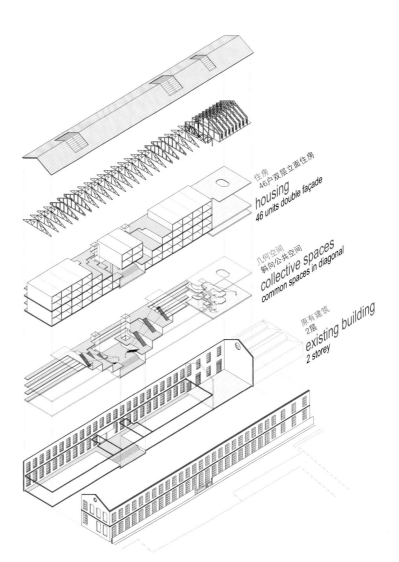

住房
46户双层立面住房
housing
46 units double façade

几何空间
斜向公共空间
collective spaces
common spaces in diagonal

原有建筑
2层
existing building
2 storey

巴塞罗那由旧工业建筑改造的文化设施与住宅单元
Cultural facilities and housing units in Barcelona converted from an old industrial building

巴塞罗那法布拉和科茨老工业建筑群的仓库改造涉及的是19~20世纪的纺织厂建筑群,改造的目的是将其纳入"BCN创意工厂"网络。该项目将为圣塔德鲁地区带来超过28 000m²的设施,而且,这也是首次在工业遗产改造中将公益住房纳入方案。该项目包括46套两居室住宅单元,其中41套供年轻人居住,其余的临时住宅单元供在建筑群内工作的艺术家居住。

改造措施激活了原有建筑的所有元素,创造了新的项目,重新利用了其物理、空间和历史特性,使新的建筑使用效率更高,同时强化了原有建筑的特性。第一个决定就是利用原建筑的最大尺寸——100m的长度。穿过中心的通道创造了一个内部广场,广场中有一部楼梯倾斜向上通往上层,跨越了两层高的空间。原建筑从一层到屋顶上下贯通,无论在实际空间上还是在视觉上都能无障碍交流。

和原来的工业建筑一样,新建筑是一种组合结构,采用只需要少量材料的干法施工,使用各种形式的木材:实心的、胶合的、交叉层压的。所有材料有机地契合在一起。结构可以组装和拆卸,它是可逆的。结构的两个内部楼层不需要任何加固件就可以支撑住房的新楼层。

建筑的外立面和屋顶为新住宅单元起到了热缓冲的作用,这些住宅单元通过一个新的木立面与建筑外立面和屋顶分隔开。空气可以在这个中间空间中流通,因此住宅单元内不再需要空调。

改造的目的是让在住宅单元中居住的年轻人和艺术家能够充分利用社交公共空间。这个公共空间是圣安德鲁区和坎法布拉内部广场之间的结合点,它把临时艺术创作空间和共享工作区合到了一起。

新建筑的设计遵循了两个原则。首先是室内外的连续性:室内空间与建筑群的公共广场相通,在广场上可以进行户外训练。其次,室内空间之间拥有视觉互动:透明的金属围合构件让人们在建筑内的任何地方都可以看到"人塔"。

"人塔"结构系统的设计基于对金字塔结构的分析。理想情况下，这种结构是纯压缩型的，它们像贝壳一样，在内部创造了一个空间。这种结构是以加泰罗尼亚的民俗节活动叠人塔命名的。在人塔中，基层需要额外的支撑来支撑顶层，基层有三层，称为平雅、福雷和马尼勒。这些设计策略可防止重量过大引起的移动或倒塌——一个楼层（或一层人）承受的荷载越多，这一层的厚度就越大，荷载的下降会显示在城堡（人塔）的形状中。这座新建筑的构思与人塔相辅相成，从结构上体现了加泰罗尼亚的当地传统。

The transformation project of the warehouse building, of the old industrial complex of Fabra & Coats in Barcelona, involves converting the 19~20th century textile complex so as to incorporate it to the "BCN creation factories" network. The project will bring more than 28,000m² of facilities to the Sant Andreu district, and, for the first time in an industrial heritage transformation, social housing is included in the scheme. The project includes 46 two-bedroom housing units: 41 units for young people, and the rest temporary residences for artists working in the complex.

The intervention activates all the elements of the original building to create the new program, reusing its physical, spatial and historical qualities to make the new construction more efficient while reinforcing the character of the original building. The first decision was to harness the value of the maximum dimension of the original building – its length of 100m. The access through the center creates an interior square where the stairs ascend diagonally across the double-height space. The original building is communicated physically and visually from the ground level up to the roof.

As with the original industrial building, the new construction is an assemblage, a dry construction with only a few materials. Wood is used in all its forms – solid, agglomerated, cross laminated; materials are joined like textiles. The new construction

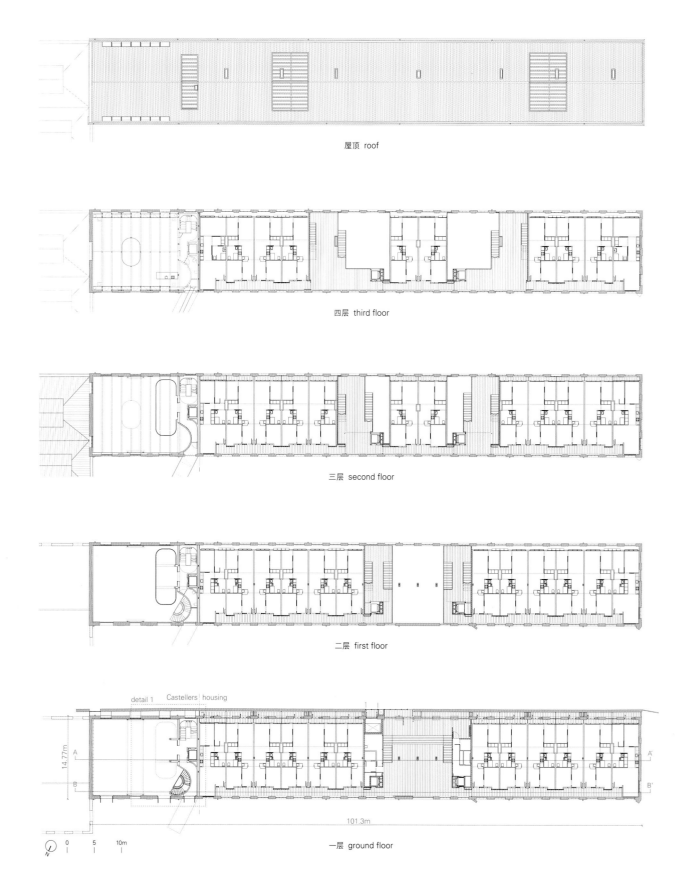

屋顶 roof

四层 third floor

三层 second floor

二层 first floor

detail 1　Castellers housing

14.77m

101.3m

一层 ground floor

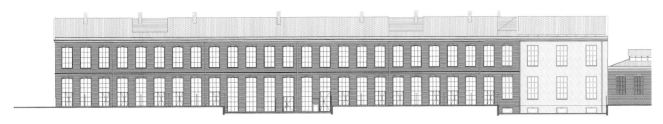

南立面 south elevation

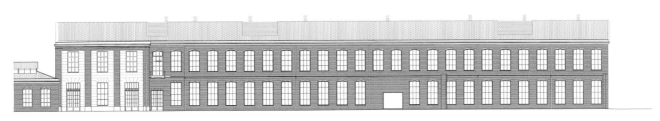

北立面 north elevation

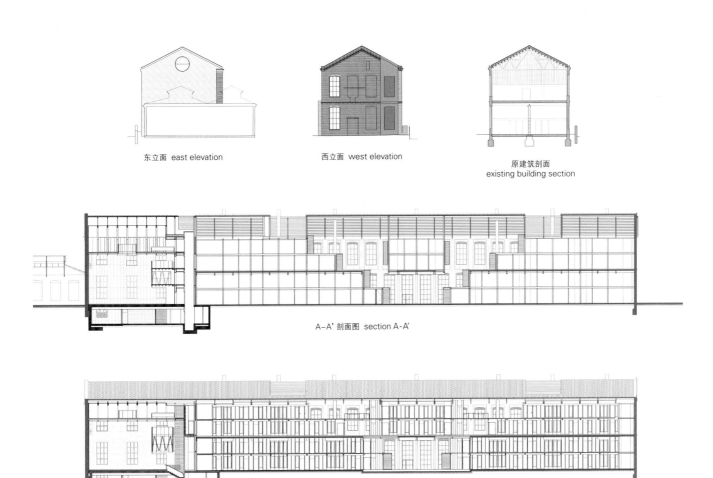

东立面 east elevation　　西立面 west elevation　　原建筑剖面 existing building section

A-A' 剖面图 section A-A'

B-B' 剖面图 section B-B'

can be assembled and disassembled, sewn and un-picked, as it is reversible. The two inner floors of it required no structural reinforcement to support the new levels of housing.

The facade and roof of the building act as a thermic buffer for the new housing units, which are separated from these with a new wooden facade. The in-between space allows air to circulate and removes the need for air conditioning in the housing units.

The intention is for the housing units to be occupied by young people and artists who are able to make use of the most part of the social common space which combines temporary art interventions and shared workspace, a point of union between the district of San Andreu and the interior square of Can Fabra. The new building follows two main principles. Firstly, interior-exterior continuity: the interior spaces open to the public square of the complex, where outdoor training can be done. Secondly, visual interaction between the interior spaces: gaps and transparent metal enclosures allow visibility of the "human towers" from anywhere in the building.

These structural systems are based on the analysis of pyramidal structures that work, ideally, with pure compression; they work like a shell, creating an empty space inside. These are named after the human towers called Castell which form part of Catalan festival tradition. In human towers, extra support from the base levels are needed to support the top layers, called pinya, folre and manilles. These strategies prevent buckling caused by movements or collapse due to excess weight – the more load a floor (or layer of people) has, the more section it has, and the drop of loads is displayed in the shape of the castle. The new structure is conceived in a complementary way to a human tower, structurally embodying a local Catalan tradition.

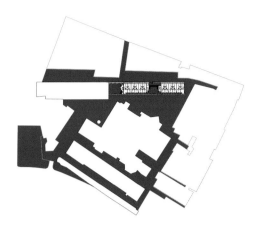

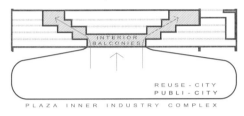

公共空间：倾斜通道通往建筑群各层
common spaces: diagonal access in relation to the complex

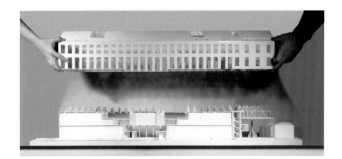

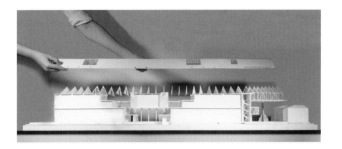

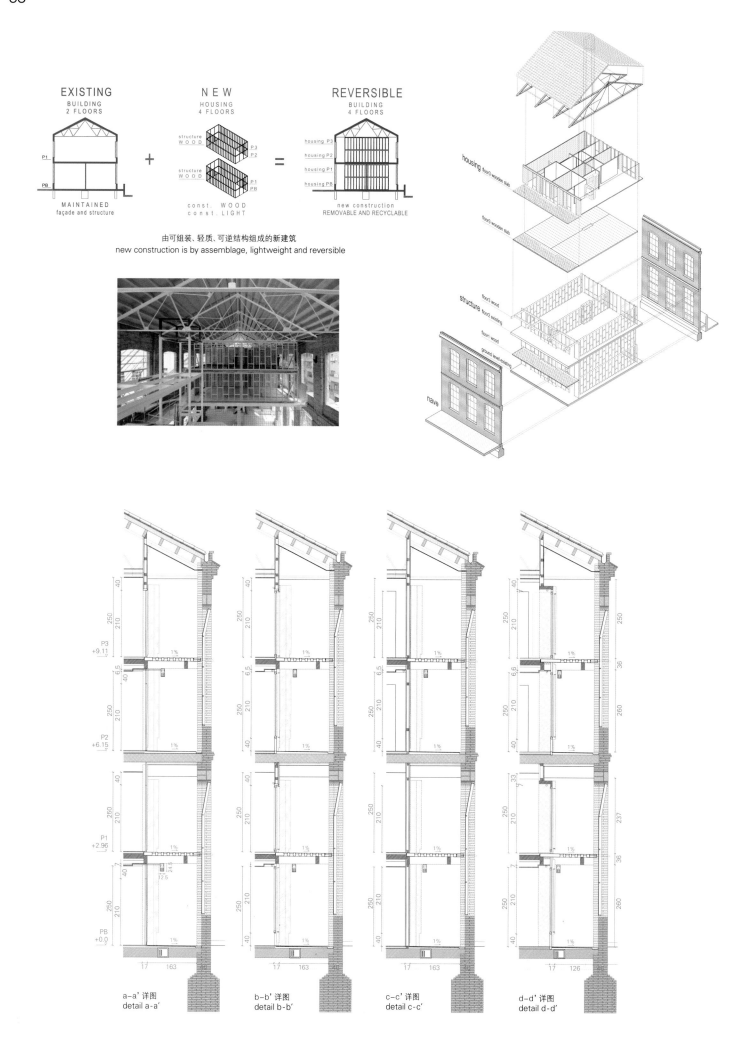

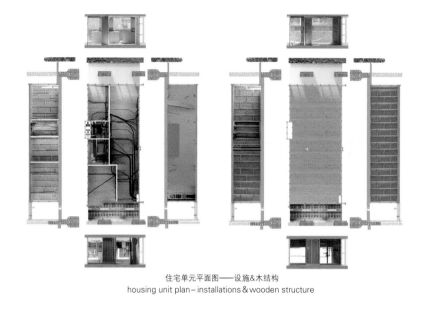

住宅单元平面图——设施&木结构
housing unit plan – installations & wooden structure

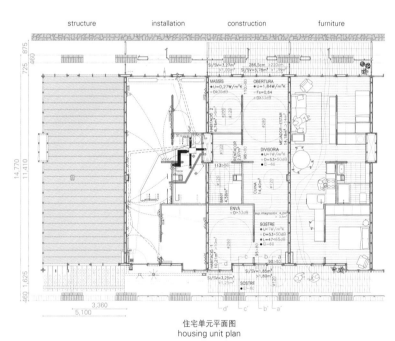

住宅单元平面图
housing unit plan

项目名称：Fabra & Coats
地点：c/ Parellada 9 08030 Barcelona, Spain
建筑师：Roldán + Berengué, arqts. – José Miguel Roldán, Mercè Berengué
合作者：Vicenç Sanz, Zana Bosnic, David Espuña, Miquel Canyellas
结构：Jordi Bernuz (Bernuz-Fernández arquitectes)
安装：Josep Dalmau (CABA)
执行总监：Salvador Arisa, Joan Just
施工单位：Sacyr
发起人：Institut Municipal de l'Habitatge, Rehabilitació, Barcelona City Council
总建筑面积：5,391.95m²
Colla Castellera建筑面积：1,113.22m²
一套住房建筑面积：59.1m²
一套住房使用面积：53.26m²
住宅单元数量：46 dwellings – 44 of 2 bedrooms and 2 of 1 bedroom
竣工时间：2019
摄影师：©Jordi Surroca + Gael del Río (courtesy of the architect)

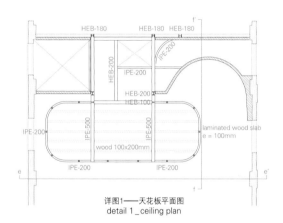

详图1——天花板平面图
detail 1 _ ceiling plan

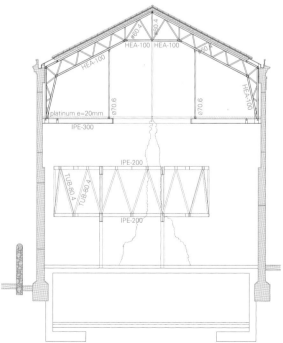

e-e' 详图 detail e-e'

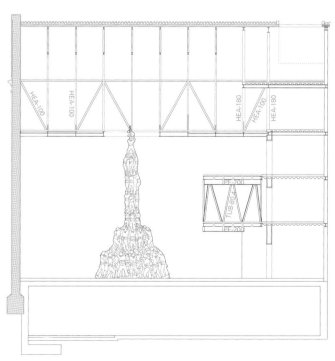

f-f' 详图 detail f-f'

将体育与科学非传统地结合在一起的英国学校
British school unconventionally combining sports and science departments in a linear volume

布莱顿学院的体育与科学系没有固守传统理念,对教育建筑的传统特征进行创新。教育建筑的传统特征是有着长长的空旷走廊和安静的教学楼。现在布莱顿学院要将这两个系的建筑合在一起,创造出充满活力的空间。在此进行的活动不仅限于学校安排的。学习过程可以在教室外进行,也可以在教室内进行,这种设计清晰地展示了能够促进互动和交流的新教育空间概念。

布莱顿学院成立于1845年,是一所私立的男女合校寄宿制和走读制学校,多年来巩固了其作为英国领先私立学校之一的地位。校园由两个部分组成:一个是历史悠久的四方形建筑,由吉尔伯特·斯科特爵士和托马斯·杰克逊爵士在19世纪设计的二级列管建筑组成;另一个是运动场,运动场两边是建于20世纪70年代和80年代的建筑,这里也是新建筑的所在地。

为什么要把科学系和体育系分开呢?在这里,两个系都被设在一个线性的建筑体量里,新的建筑就设在运动场的旁边。主要的运动空间与运动场相邻,体育馆直通运动场。科学系包括教室、实验室和温室,这些空间就像一座桥梁骨架横跨在运动空间上方。建筑立面设计的灵感有一部分来自于新建筑对面的联排住宅。在屋顶上可以俯瞰英吉利海峡,视野非常开阔。

激发学生之间的交流是这一设计理念的核心:教室外宽敞开放的休息区域为非正式活动和个人学习提供了空间。楼层的改变、宏伟的楼梯和玻璃在视觉上把两个系的活动联系起来,并促进了两个系不同学科的交流。建筑的各个组成部分相互可见:从上层可以看到一层的室内跑道,教室有落地窗,甚至化学教室里的通风柜也是透明的,这样人们就可以在走廊看到正在进行的实验。

布莱顿学院体育与科学系大楼
Brighton College Sports and Science Building

OMA/Ellen van Loon

OMA在2013年的建筑设计方案竞赛中赢得了这个项目，当时布莱顿学院正准备扩建科学系实验室，增加先进的体育设施，实现培养人才、促进师生身体健康的目标。在最初的设计竞赛任务书中，运动中心和科学系大楼是两个独立的项目，2013年修改任务书，2014年又举办了一次设计竞赛，2015年计划获批，2017年开始建设。该项目总预算为3670万英镑。

The School of Sports and Science at Brighton College defies the conventional character of educational buildings – one of the endless empty hallways and imposed silence – and instead combines the two departments to create a vibrant building with lively spaces where activities are not necessarily dictated by a school timetable. Observing that processes of learning take place outside as much as inside of the classroom, the design articulates a new idea of educational space bolstering interaction and exchange.

Established in 1845, Brighton College is a private, co-ed boarding and day school in Brighton, England, and over the years has cemented its reputation as one of Britain's leading independent schools. The campus is comprised of two areas: a historical quadrangle, composed of Grade II listed buildings designed by Sir Gilbert Scott and Sir Thomas Jackson in the 19th century; and the playing field lined with buildings from the 1970s and 1980s, the site of the new building.

Why does the design isolate the department of Science from

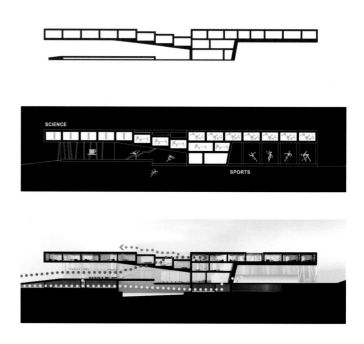

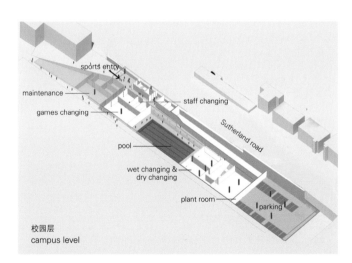

校园层
campus level

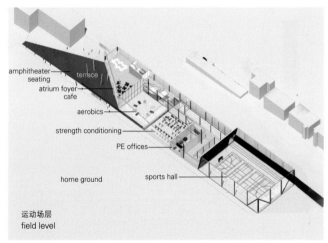

运动场层
field level

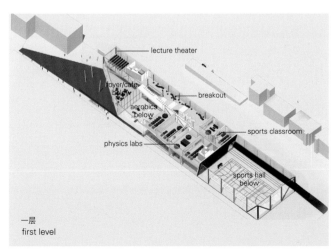

一层
first level

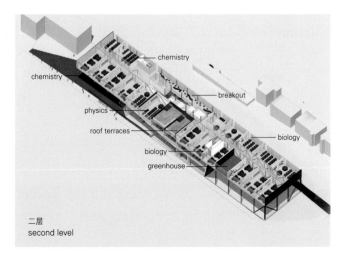

二层
second level

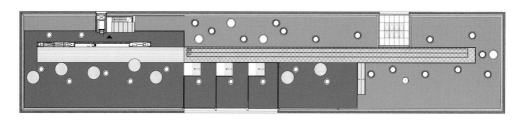

屋顶 roof

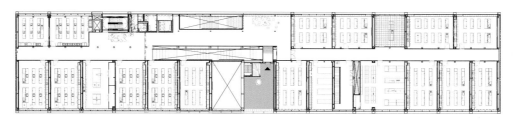

二层 second level

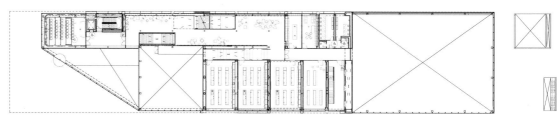

一层 first level

运动场层 field level

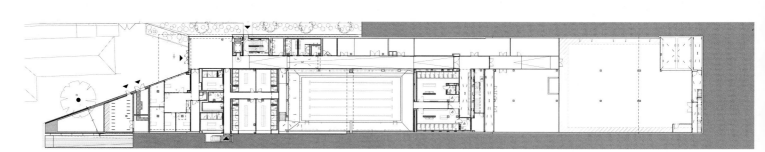

校园层 campus level

项目名称：Brighton College Sports and Sciences Building / 地点：Eastern Rd, Brighton BN2 OAL, UK / 建筑师：OMA–Ellen van Loon / 项目总监：Carol Patterson / 设备工程：schematic design–Carol Patterson, Marina Cogliani, Cristina Marin de Juan, Dinka Beglerbegovic; design development–Carol Patterson, Marina Cogliani, Michalis Hadjistyllis, Mario Rodriguez; construction–Carol Patterson, Isabel da Silva, Marina Cogliani, Tom Pailing, Magdalena Stanescu / 结构工程：Skelley and Couch / Structural engineering: Fluid Engineering / 景观：Bradley-Hole Schoenaich / 声学：Ramboll / 消防工程：The Fire Surgery / 可持续发展：Eight Associates / 承包商：McLaren / 雇主代表：Gardiner & Theobald / 客户：Brighton College / 功能：educational (science class rooms, laboratories, sports facilities, swimming pool) / 总楼面面积：7,425m² / 预算：£36,700,000 / 竞赛时间：2013~2014 / 施工时间：2017~2019 / 竣工时间：2020 / 摄影师：©Laurian Ghinitoiu (courtesy of the architect)–p.50~51, p.52, p.53, p.57[upper], p.58~59[left-upper], p.62, p.63; ©Killian O_Sullivan (courtesy of the architect)–p.48~49, p.51[lower], p.52~53[left-lower, right], p.54, p.55

the department of Sports? Here, the two are merged into one linear volume, situated at the edge of the playing field. Primary sporting spaces are level with the field, and the sports hall opens directly onto it. The Science department, which include classrooms, laboratories and a greenhouse, spans over the sporting spaces like a skeletal bridge. The facades are inspired in part by the regular rhythm of the terraced housing opposite the new building. The rooftop provides a sweeping view of the English Channel.

Stimulating social communications between the students was central to the concept: generous and open break-out space outside of the classrooms create room for informal interaction and private studying. Level shifts, grand staircases and glass visually connect the activities taking place in both departments and trigger unexpected exchanges between different disciplines. Individual components of the building are exposed to each other: an indoor running track on the ground floor is visible from upper levels; classrooms have floor-to-ceiling windows; even fume hoods in the chemistry classrooms are made transparent – enabling people walking down the hallway to witness ongoing experiments.

OMA was awarded the project after a competition organized in 2013, when Brighton College needed to expand the Science School in terms of number of labs, and wanted a state-of-the-art sporting facility to foster talent and physical wellbeing. In the original competition brief the Sports Center and Science Department were presented as two separate projects. After a revised brief in 2013, and a second competition phase in 2014, planning approval was acquired in 2015 and construction started in 2017. The project was realized with a total construction budget of £36,700,000.

A-A' 剖面图 section A-A'

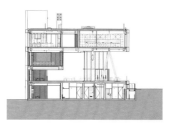

B-B' 剖面图 section B-B'

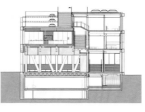

C-C' 剖面图 section C-C'

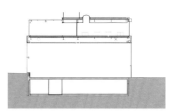

D-D' 剖面图 section D-D'

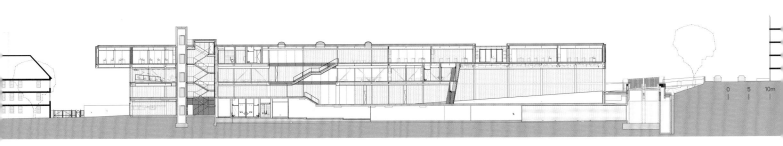

E-E' 剖面图 section E-E'

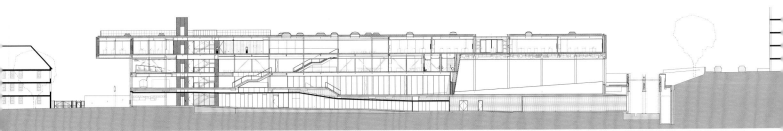

F-F' 剖面图 section F-F'

保持社交距离，守护城市密度

Defendin
in the Year of S

通过这一部分列举的四个项目，我们将为您展示多种既不扰乱城市社会生活，同时又能达到合理城市密度的方法。城市密度过高，百姓彼此难免心生提防，造成公民生活日渐衰退；而城市密度低疏，街坊之间距离又太远，往往会不相往来。解决问题的题眼就在于"度"。城市密度得当，市民不仅不会产生陌生感，还能避免不得不闯入他人生活的尴尬。除此之外，打造浓郁城市生活感的另一种方式，就是实现住宅、娱乐、商务等用途的集成。本文所举四个项目均为9～11

The four projects featured in this section show a variety of ways of achieving density while maintaining the social life of the city. When a city is too dense, it becomes alienating, and civic life declines; when the density is low, people tend not to share aspects of their lives. The important question is scale. The proper scale of density allows residents to become aware of each other without necessarily knowing each other. To mix housing with work, play, and commerce is the other method of creating a strong sense of urban life. All of

Carlo Erba住宅楼_Carlo Erba Residence/Eisenman Architects + Degli Esposti Architetti + AZstudio

Fenix一期项目_Fenix I/Mei architects and planners

Bottière Chénaie住宅综合体_Bottière Chénaie Housing Complex/KAAN Architecten

Denizen Bushwick公寓楼_Denizen Bushwick/ODA

保持社交距离，守护城市密度_Defending Density in the Year of Social Distancing/Richard Ingersoll

Density
ocial Distancing

层建筑，分别位于纽约布鲁克林、南特、鹿特丹和米兰。它们营造了热闹的街景，提供了宽裕的开放空间，带来了充分的服务便利。其中两个项目加入了魅力十足的工作空间，为社交空间带来了生机活力。有观点认为，可持续发展不仅要求我们在技术上减少能源用量，更重要的是，它为我们指明了社会前景。在本文的四个项目中，这一前景得到了成功实现。

the featured projects in Brooklyn, Nantes, Rotterdam, and Milan stay within a 9-11 story volume and attempt to make active streetscapes. They also offer ample open spaces and access to services. In two cases the projects contain intriguing work spaces that enliven the social spaces. If sustainability requires not only technical reduction of energy use, perhaps more importantly sustainability is a social prospect, one that in these projects has been successfully implemented.

保持社交距离，守护城市密度
Defending Density in the Year of Social Distancing

Richard Ingersoll

  总的说来，打造可持续城市的重点工作之一是提高城市密度。大型公寓建筑能够大幅度减少能耗，降低用水需求，缩小人均占地，同时推广多项共享便民设施；交通站点附近的紧凑型住宅可以大大降低人们对机动车交通的依赖程度。但面对新冠肺炎疫情，我们不得不重新考虑：市民应当比邻而居的信条，是否仍旧正确？密度越大，感染概率是否就会越高？毕竟与独门独户，房有后院的人相比，这段时间宅在公寓中的人们接触大自然的机会更少，感染风险更大，特别是在规模大、污染重的都市中，风险尤甚。不过无论如何，我们都期盼人类社会能够最终战胜病毒，纵使疫情反复，也愿广大群众能够准备得当，应对自如。

  但现在，我们还有更严重的环境危机要克服。如果我们回归高熵值的消费主义生活方式，全球变暖将会大大加重。因此，即便城市密度无法迟滞病毒传播，它仍是值得我们重视，特别是在新冠肺炎时代后必须践行的一条理念。Peter Calthorpe和其他城市学家一致认为，解决二氧化碳聚集问题的一大障碍就是美式城市摊铺布局，该布局由低密度定居点构成，且定居点多为独门独户的房屋。虽然目前看来，这种居住方式在遏制感染方面似乎较为安全，但在能源消耗和交通运输方面，则会产生有害的影响。作为这种美式布局的反面，纽约市就是一个很好的例证。纽约市的城市密度全球第一，只有市长才是独门独户。尽管这使得纽约市成为在疫情中损失最惨重的城市之一，但在曼哈顿地区，保持城市密度却能自动减少集体碳足迹。本文将通过四个住宅项目，为大家进一步提供有关城市密度的新视角。这四个项目并非追求容积率的最大化，而是提供了大量的开放空间，便于市民欣赏自然景致，为共居方式打开了新的思路。其中，布鲁克林Denizhen Bushiwick项目打破了美式摊铺布局（见118页），为我们提供了极为重要的例子。项目

By all accounts, one of the keys to creating sustainable cities is to increase density. Living in large apartment buildings can significantly reduce energy and water needs and the amount of land per person, while promoting many shared amenities. Compact dwellings near transit stops obviously minimize one's dependence on the automobile for transportation. Alas, the COVID-19 pandemic forces us to reconsider the virtues of living together in such proximity. Does more density lead to greater chances of contagion? Certainly those who have been confined to their apartments during this period have not only had less opportunities for contact with nature than those in a house with a backyard, but also have been more exposed to contagion, especially if living in a large, polluted metropolis. We all expect, however, that the virus will eventually be brought under control, and if it returns, hopefully urban dwellers will be better prepared to resist it.

The greater environmental crisis, however, that of Global Warming, will definitely worsen if people return to live high-entropy consumerist life styles. So while density may not help to deter the spread of a virus, it must be respected as part of the Post-Corona mandate. As Peter Calthorpe and other urban theorists agree, one of the prime enemies in the struggle against CO_2 build up is American-style sprawl, based on low-density settlements of single-family houses. Alas, at the moment, such dwellings seem safer in terms of contagion, but energy-wise and transportation-wise, they prove to be deleterious. New York City provides the great counterpoint for the United States, one of the densest cities on earth – no one besides the mayor lives in a detached residence. Alas, due to its extreme density, it has been one of the hardest hit by the virus. Still Manhattan demonstrates how density automatically lowers the collective carbon footprint. The four housing projects in this review propose a different attitude to Manhattan's density. Rather than maximizing FAR, they provide lots of open space, contact with natural elements, and indeed imply new ways of living together. The most important in terms of rejecting American sprawl can be seen in the Denizhen Bushiwick in Brooklyn (p.118). It occupies two of the ten blocks that once housed the Rheingold

垂直森林，意大利，2014年
Vertical Forest, Italy, 2014

以社区建设为导向，性质为混用住宅，占用了原Rheingold酿酒厂十座大楼中的两座。大楼屋顶的空中花园壮秀华美，草木郁郁葱葱，整个项目更是出类拔萃。在Fenix一期项目中，前工业的风格也起到了很好的效果（见82页）。项目位于鹿特丹的莱茵港码头，在这片早已废弃的荷美航线船坞上，新建大楼拔地而起，与Denizhen Bushiwick相似的是，Fenix一期也是多元化社区用途和公共空间项目。另外两个项目同样是舒适城市密度的范例。在选址决策时，它们并未过多考虑社区需求，但仍为社区建设提出了坚实可靠、毫不敷衍的解决方案。

还有许多住宅项目的设计方案堪称是才华与汗水的结晶，但在建筑历史中，它们却鲜有留名，这多少让人感到意外。科布西耶设计的马赛Unité建筑、密斯设计的芝加哥湖畔公寓、伍重设计的集群住宅以及其他卓尔不群的作品，例如，黑川纪章设计的东京中银胶囊塔楼和Lucien Kroll设计的布鲁塞尔郊外La Mémé学生公寓，都是这方面的范例。住宅项目缺席建筑历史的原因有很多：一方面是历史偏好，人们更看重具有纪念意义的建筑；另一方面是分野不同，住宅似乎属于社会学等学科。纵观现代，只有在类型学方面实现重大突破创新的住宅项目才有希望载入史册，如BIG在哥本哈根设计的山形住宅、Boeri在米兰设计的垂直森林以及Lacaton & Vassal在法国设计的包裹旧公共住宅结构的暖房等。不过，住宅项目的舒适度、功能规划以及可持续发展等因素仍是我们需要考虑的重点，特别是在目前，情况尤为如此。例如，在曼哈顿、香港、新加坡等用地有限的地区，仍有棘手难题亟须破解：解决城市密度问题，应当增加建筑高度，还是改进大楼包装效果？出于各种原因，本文所举项目提出了9~11层建筑解决方案。这些项目或是退让空间，或是采用阳台，或是穿墙搭桥，有效利用了街道元素，大大厘清了建筑与街道和公共空间的关系。事实证明，

brewery and now carries an exceptional, community-oriented mixed-use housing project, densely packed and capped with a magnificent rooftop garden. Its ex-industrial demeanor goes well with the Fenix I (p.82) project on the Rijnhaven pier in Rotterdam, which heroically rises over of the abandoned docks of the Holland-America shipping lines. It likewise offers a diverse program of community functions and public access. The other two projects presented herein, both models of comfortable density, are on sites less determined by the demands of community organizations, yet demonstrate strong formal solutions in the effort to correspond to a community.
It seems odd that so few housing projects ever break into the ranks of architectural history considering how much talent and energy go into their design. We will find Corbu's Unité in Marseilles, Mies' Lakefront apartments in Chicago, Utzon's cluster dwellings, and some eccentric works such as Kisho Kurokawa's Nagakin Capsule Tower in Tokyo and Lucien Kroll's La Mémé student housing outside of Brussels. This oversight is partly due to the historical bias for monuments, but also because the dwelling seems to be a subject belonging to a different discourse, such as sociology. In general only the projects of significant typological innovation, such as BIG's Mountain Dwellings in Copenhagen, Boeri's Vertical Forest in Milan, or Lacaton & Vassal's greenhouses wrapping the old public housing structures of France, have recently shown the sort of appeal that will extend beyond the realm of current events. Yet the factors of comfort, program, and sustainability of housing, especially now, need our attention, and the following examples, while they will not become canonical in architectural history, are exemplary densifiers.
The big question, which is difficult to ask in places with a limit of land like Manhattan, Hong Kong, or Singapore, remains: should density be achieved through greater height or through better packing of the block? For various reasons the featured apartments herein have pursued solutions from 9 to 11 stories, negotiating a clear relationship with the street and public space. This effort to integrate with the city block structure proves quite different than the strong alienation from the urban context found in most tall

Bottière Chénaie住宅综合体，法国
Bottière Chénaie Housing Complex, France

与大部分高楼大厦在城区中造成的突兀隔绝相比，融合城市街区结构起到的效果大有不同。

在这四个项目中，布鲁克林Denizhen Bushiwick项目的历史最为复杂。该项目紧紧围绕社区理念展开，力图在贫乏不振的周边环境中、在已被污染的棕地上建立一片公平合理的住宅。项目原址为Rheingold酿酒厂，已于1981年拆除，留下的只是一片荒凉毒土，需要环境保护机构下大力气去除污染；同时，当地政府还想将这片土地打造成公屋住宅，供弱势群体和少数族裔居住。2000年，该区域部分地块被开发为保障性住房。当时的开发商发现附近Williamsburg的创意产业住户人丁日盛，认为需求有所变化，遂决定将2~3层建筑扩建为9~11层大楼。不过，鉴于项目与当地社区息息相关，居民深恐中产阶级移居侵占，为此Eram Chen带领的ODA建筑师团队在设计时不得不做出让步，将部分项目改为向社区开放，同时提供了20%的保障性住房。整体来看，Rheingold项目的外立面有着鲜明的斜交网格结构。内部公共街道和各式院落点缀着项目的两栋大楼，植被茂密，楼体上涂满了巨大的涂鸦壁画。项目功能多样，让人想起大学校园，这对刚出校门的职场新人具有很强的吸引力。项目设有酒吧、餐厅、室内泳池、拳击台、健身房、保龄球室、少儿高尔夫球场、微型酿酒厂、木工实验室、金属加工实验室、视频制作与摄影师空间、展览室、儿童游乐场以及设有壁炉的社区休息厅。大楼屋顶的空中花园草木繁盛，还有城市农场可供休闲，提供了多姿多彩的社交空间。居民可在此凭栏远望布鲁克林大桥和曼哈顿，景致之妙，远超勒·柯布西耶承诺的同类屋顶花园设计。在这些设计特点的支持下，项目为3000名居民打造了极其活跃的社交空间。目前，新冠疫情肆虐，社交需要严格监管。这里的居民彼此熟识，形成了贯穿项目的社区价值观，有助于进一步提高责任感，做好疫情防控工作。庭院、屋顶和阳台上种植名花异草，

buildings. In different ways, each of these projects plays with the street through stepping back, terracing over, and poking through.

The Denizhen Bushiwick in Brooklyn has by far the most complicated history, tied to a community driven effort in a poor neighborhood to obtain equitable housing on a polluted brownfield site. The Rheingold brewery was demolished in 1981, and the site required immense bureaucratic efforts by the Environmental Protection Agency to eliminate toxins, while local politicians sought to promote fair housing for a disadvantaged, ethnic neighborhood. In 2000, some of the area was developed as affordable townhouses. The developers of the project under review, sensed a change in demand, due to the increasingly creative settlers in nearby Williamsburg, and decided to scale up from two- and three-story buildings to large blocks of nine stories. The community held a strong stake in the project, fearing gentrification, and when ODA architects, led by Eram Chen, took on the design they were obliged to make some of the project open to the community while providing 20% affordable units. The exterior facades of the Rheingold display an impressive sheer diagrid structure on a grand scale. The two blocks are penetrated by an internal public street and variegated courtyards, intensely planted and extravagantly decorated with gigantic murals. The exceptional variety of functions included in the program is reminiscent of a college campus and indeed appeals to young professionals just out of college. They include bars and restaurants, an indoor swimming pool, a boxing ring, a gym, a bowling alley, a pee-wee golf course, a micro-brewery, laboratories for woodworking and metal working, spaces for video-makers and photographers, exhibition rooms, a toddlers' playground, and community lounges with fire places. The intensely planted roofscape, part of which is an urban farm, offers a variety of social spaces that look to the Brooklyn Bridge and Manhattan beyond, going far beyond Le Corbusier's promise of rooftop gardens. All of these features exude a certain program of social promiscuity for the 3,000 residents, and in these times will require strict supervision. But one can imagine that the community values imbedded in the project will probably lead to a better than

Fenix一期项目，荷兰
Fenix I, the Netherlands

涂鸦壁画泼洒鲜活色彩，内部风格采用时髦的混搭，大厅中央树干拔地而起，直通上一层，整个项目散发着狂野不羁的气息，避免了风格一致造成的压抑感。

但KAAN建筑师事务所的作品——法国南特Bottiére Chénaie住宅综合体（见98页）维持城市密度的方式方法却和上例截然不同。该项目严格遵循现实主义风格，对功能混搭进行了严密的限定。这一严谨正式的态度也构成了密斯公司空间句法的基础，最近推出的Diener & Diener等办公建筑就是范例。Bottiére Chénaie项目成功采用预制混凝土框架等重装材料，配以优美细节装饰，不仅颇有安藤忠雄的精准，还散发着熠熠生辉的透明感，轻盈效果远超大部分功能主义作品。项目位于南特城东北郊的普通区域，公共交通非常便利，但高速公路基建略多。为此，KAAN选择远离街道，搭建了更为清净的内庭院。项目三栋大楼各有用途，分别是办公楼、市场价住房和补贴住房，坐落于同一个三层基座上。在平等要求方面，KAAN并未被社区组织者左右，而是以政府政策为准绳，最终正式达成了"项目用户无预定社会群体"的目标。住宅楼由人行通道连接，阳台空间宽敞，经过精心设计，保留了开放空间的设定，是活跃社交活动的好去处。办公室和172套公寓的用户可选择基座内的地下停车场，不必沿街停靠车辆。除此之外，基座内还设有商务区域。不过项目距离当地商场很近，因此购物区并不需要很大。办公室的垂直框架和住宅的水平框架都采用了淡素静美的混凝土色系，纵横搭配优雅非常。这片土地原本平凡无奇，但在KAAN的精心灌溉下，现已是欣欣向荣的沃土。

Fenix一期公寓项目由鹿特丹的Mei建筑规划公司设计。项目没有因循守旧，转而向MVRDV的新荷兰方式靠拢，由

average sense of responsibility, since neighbors are quite aware of each other. The luxurious planting of the courtyards, roofs, and terraces, combined with the exuberant colors of the murals and the funky mix of interior styles, including a lobby with a tree trunk pushing through the bottom floor to the next story creates a rather wild atmosphere but alas does not lead to any compelling stylistic coherence.
Such could not be farther from the approach of KAAN Architects at the Bottiére Chènaie Housing Complex (p.98) project in Nantes, where a severe sachlichkeit style regiments the mixture of functions. The formal attitude evokes the syntax of Mies, seen more recently in the works of offices like Diener & Diener. The beautifully detailed precast concrete frames recall the precision of Tadao Ando's work. The success of using such heavy materials to obtain such luminous transparency gives the project a feeling of lightness that most functionalist projects never achieve. The site of the project is in a nondescript northeastern suburb of the city with good access to public transportation. The area is a bit overwhelmed by highway infrastructures, which explains the project turns away from the street to provide more tranquil interior courtyards. The three block-size buildings, serving three different programs for offices, market rate housing, and subsidized housing, are connected by a three-level plinth. In this case, it was government policy and not neighborhood organizers that led to the demand for equity, and the formal result speaks to this vision of a place without a pre-existing social base. The generous balconies and connecting walkways between the housing blocks promise to enliven social interaction and have been carefully designed to remain as open spaces. The plinth provides below grade parking for the offices and the 172 apartments, keeping the cars off the street, while part of the ground floor contains commercial areas, but since the project is quite close to a shopping mall, there was no great demand for extensive shopping areas. The vertical frames of the office structure and the horizontal of the housing provide an extremely elegant combination in a sober concrete palette. Kees Kaan has instilled this featureless territory with a clear generative language.
Mei architects and planners, another office from Rotterdam designed the Fenix I lofts, immune to a

Carlo Erba住宅楼，意大利
Carlo Erba Residence, Italy

表及里，将功能性做到极致。建筑师面对的第一个问题是：如何大幅度增加荷美码头这一历史建筑的空间，并在保存原有建筑的同时启用新基础？Mei公司从原建筑混凝土结构上的寄生式钢吊车中汲取灵感，提出了自己的解决方案。设计人员首先做成百万吨的钢"台"，再在台顶安装一个4m见方的钢框架，支撑建筑物上层；然后将框架形成的中间层分割开来，安上玻璃，形成公寓的第一层；最后引入混凝土通道解决方案，在框架层上构筑8层建筑，将墙体和楼板浇筑为一个整体。在这8层建筑中，下3层为出租单元，而上5层的130个单元则由Mei公司和单元业主互助自定义设计而成。在北侧靠码头一侧，外立面共有11层，与周边建筑规模相仿；南侧外立面只有6层，整体建筑呈现向现有街道方向逐渐缩小的态势。公寓没有采用简单的平行六面体外形，南侧较矮一侧呈斜坡状，便于阳光照耀空间宽敞的庭院；北侧较高一侧则几乎为玻璃墙，2.5m宽的阳台环绕着玻璃墙，整个结构以"缪斯"框架为基础。公寓内里光亮辉煌。底层的原有结构被分割开来，形成了一个停车场和宽敞的实验室空间；实验室空间又分为三个教学区域，分别为舞蹈学校、马戏学校和烹饪学校。一条玻璃通道从庭院中央穿过，公众可在其中徜徉，一睹绿植阳台，享受绿荫遮蔽。除此之外，更有多家酒吧、餐厅紧邻码头和项目边沿排列。Fenix一期公寓项目与布鲁克林的Rheingold项目类似，城市密度经过调整，适应了社交需求。在社交距离相关措施方面，该项目也可能起到集思广益的正面作用。

　　本文要介绍的最后一个项目是Carlo Erba住宅楼（见66页），位于米兰，由Peter Eisenman与当地建筑师Lorenzo Degli Esposti联手推出，期间意大利设计师Guido Zuliani在纽约也助力不少。过去50年内，Eisenman以一己之力，反功能主义设计而行之，依据几何排列和地形特点，构建了全球最为精灵古怪的建筑。不过，也许是此次合作的设计师

preconceived language and more in line with the New Dutch method of MVRDV, letting functions take form to the extreme. The first question: how to make a significant volumetric addition to the historic Holland-America dock building, saving the original but using new foundations. The solution was inspired by the parasitical steel cranes that once rose on top of the original concrete structure, resulting in a one-million-kg "table" of steel, capped with a 4m-space frame to support the upper levels. The designers subdivided and glazed this intermediate level for the first layer of lofts. They then added a concrete tunnel solution to build the upper eight layers that rise above the steel frame so that walls and floors could be cast together. Here the first three levels serve rental units, while the remaining upper levels of 130 units were custom designed with the collaboration of each owner. On the northern dockside the facade has 11 stories in keeping with the scale of nearby buildings while on the south it only has 6 stories to scale down to the existing street. Thus instead of a simple parallelepiped, the short side leaves a sloping profile, which of course allows the sun to enter into the generous courtyard. On the tall side, the facades are nearly completely glazed with 2.5m-wide balconies ringed with glazed parapets, structured on "muse" frames. The interior of the lofts appear incredibly luminous. The original structure on ground level has been subdivided into a parking garage and vast laboratory spaces three cultural institutions – a dance school, a circus school, and a culinary school. A glazed passage way runs through the middle of the courtyard open to the public revealing the planted balconies providing a green shelter, while bars and restaurants line the dockside and the edges of the project. Somewhat like the Rheingold in Brooklyn the density has been geared to socializing, but this may have a positive effect in brain-storming for social distancing practices. The final project in Milan, the Carlo Erba Residence (p.66), comes from Peter Eisenman, who joined forces with local architect Lorenzo Degli Esposti. Another Italian, Guido Zuliani, collaborated in New York. During the past five decades Eisenman has led a personal campaign against function-based design, following the

较为年轻，整个项目实际上较为实用，并未凸显Eisenman建筑常见的特点。在这一点上，我们也要感谢Esienman的偶像Giuseppe Terragni，他于20世纪30年代在米兰设计的Casa Rustici公寓就是本次Carlo Erba住宅楼项目的原型。设计师从中汲取灵感，通过框架开放大楼，形成透明效果。Carlo Erba住宅楼与米兰教育机构Politecnico园仅隔三个街区，俯瞰Carlo Erba广场。广场苍翠葱茏，设有便民电车站。20世纪70年代以前，住宅大楼为出版公司Rizzoli的总部驻地，之后又成为苏黎世保险公司的办公楼。大楼经历损毁变迁，目前只剩下20世纪初期的部分后现代结构，面对着Carlo Erba广场。设计师没有按照大楼的长方外形补全剩余空间，而是保留空心位置，并在此开辟了一个庭院，从而将整个项目扭转为巨大的S形，与米兰大多数笔挺垂直的高层建筑形成强烈对比。大楼下三层按水平方向延展铺开，外立面以石灰华覆盖，这一区域不仅是大楼的入门区，也是大楼在历史变迁后留下的残余；第四层嵌入下面的楼层，是整栋建筑透明的夹层；五层、六层则采用了陶瓷金属板外墙；七层、八层和九层采用了开放式空间设计，通过大型藤架搭建了"城中别墅"专享的奢华市场和郁郁葱葱的天井露台。建筑的S形大拐弯处俯瞰着街道对面的公园，为庭院中的草坪喷泉斜角小花园遮风挡雨。建筑上层的高大框架勾勒出梦幻一般的外形，使Carlo Erba住宅楼卓尔不群，更添开放色彩。

归根结底，有关城市密度的问题，考验问的是人们群居的质量，而非住宅的数量。本文所述的四个项目在维持高度开放、紧密拥抱自然和可持续发展社交空间方面做出了大量努力，有助于提升责任社区的自觉度，有助于推进居民之间的合作，有助于解决眼下的危机和未来的挑战。

dictates of geometric permutations and topographic indicators, leading to some of the most impractical buildings in the world. Perhaps due to his young collaborators this project actually works fairly well and apparently does not have the famous leaks of so many of the theorist's built works. We also can thank Eisenman's hero Giuseppe Terragni, whose Casa Rustici apartments in Milan, built in the 1930s, provided a prototype of how to open the block with frames to create transparency. The site is three blocks away from the Politecnico campus, the city's major educational institution, and faces a small verdant plaza, Piazza Carlo Erba, where there is a tram stop. The large block, which until the 1970s was the headquarters of Rizzoli publishers and later was used as offices for Zurich insurance, was scraped leaving only part of an early 20th century neoclassical structure facing the piazza. Instead of filling out the rest of the triangular block according to its contours, leaving a courtyard in the center, the project evolved into a gigantic "S" shape that struggles against the orthogonality of most Milanese blocks. Arranged in horizontal bands, the first three stories, clad in travertine, refer to the scale of the historic fragment left as the entry. The next floor, inset from the lower levels, has been handled as a transparent mezzanine. Above this are two stories clad in enamel metal panels. The top three levels have been eroded with grand pergola frames serving the luxury market for "urban villas" with extraordinary planted patios. The largest curve of the serpentine volume, overlooks the public park across the street and shelters a small private garden of lawns and fountains composed with a series of oblique. The towering frames that outline phantom volumes on the upper stories give the Carlo Erba condominiums a heroic look of difference and openness.

The question about density in the end is certainly more about how we live together than quantities. In each of these projects the effort to maintain a high level of openness, good connections to nature, and viable social spaces will promote a certain conscientiousness for a responsible community, ready to cooperate in confronting the immediate crisis as well as those of the future.

Carlo Erba住宅楼
Carlo Erba Residence

Eisenman Architects + Degli Esposti Architetti + AZstudio

S形四阶住宅建筑，呼应米兰城市文脉
S-shaped residential building with four different layers responds to the Milan urban context

该项目地块呈三角形，周边建筑普通平凡，缺乏特点。项目设计者试图打造一栋标志性的当代住宅公寓，并在其中设置多个商用单元，与米兰城市文脉特别是周边公园形成呼应。为此，设计方还将公园扩展至项目地块处。受高度限制，建筑师依据当地法规，设计了一系列水平带，并在垂直叠加上略有偏移，形成了这栋四阶九层建筑。

建筑的一、二、三层是建筑的基座层，它们类似米兰历史悠久的城市广场，立面覆盖石灰华，并设有大小相近的打孔式窗户和嵌入式阳台。建筑的四层为第二阶，类似于传统的主厅，与下面的石灰华立面和上面的大理石立面相比，第二阶要向内缩进一些，所采用的玻璃装饰与建筑的石材表面也形成了鲜明对比。第三阶包括五层和六层，特点是采用陶瓷金属框架。框架形成了水平带，横跨外立面的整个长度方向。框架在建筑西侧设在真正的西立面前面，而东立面的框架则构成具有雕塑感的格栅，占据了建筑实体的一部分。第四阶即最上一阶，分别是七、八、九层，在某种意义上来说，它们并不构成水平分层，它们的布局和体量采用了台阶式设计，形成了一系列带有郁郁葱葱的露台的"城市别墅"。四个阶层虽然各不相同，但却搭配得当，略微偏离几何重心，极富动感；加上传统材质和窗户开口方式，堪称是今日米兰的标志之作。

相关设计方经过多次研讨，才确定了建筑的S外形。如示意图所示，该外形不仅符合功能和区划要求，还保留了地块上的20世纪早期经典建筑，将其融入公寓之中，并将其作为公寓的正式入口。建筑西侧的公园经过扩展，延伸至了公寓西立面的曲线处。

For a triangular site bounded by non-descript buildings, this project attempts to create a symbolic contemporary apartment building with the number of units required for commercial viability while responding to the context. In particular, it aims to respond to the adjacent public garden, by continuing the garden onto the site. Restricted by the height limit, due to the local regulations, the architects designed a series of horizontal bands stacked with slight offsets that create four different layers in the nine-story building.

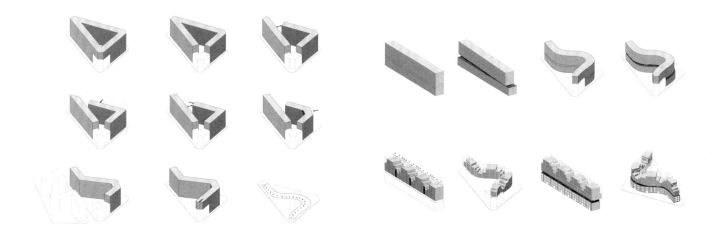

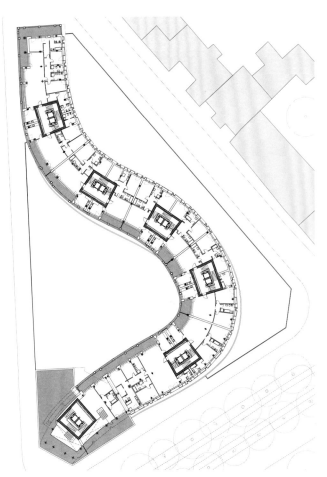

六层 fifth floor

四层 third floor

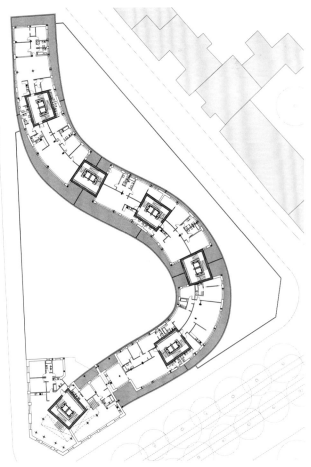

五层 fourth floor

The first three floors constitute a base course similar to historic urban palazzi with travertine cladding, punched by window openings of similar size, and inset balconies. The fourth floor, or second layer, is like a traditional piano nobile; it is set back from both the face of the travertine base and the marble face above, and its glazing contrasts with the stone surfaces. The third layer, floors five and six, is articulated by enameled metal frames that create a horizontal band running the length of the facade. These frames are set forward of the actual west face of the building. On the east facade, they excavate the mass of the building in a carving lattice. The fourth and topmost layer, floors seven through nine, is in one sense not a horizontal layer but, in its stepped profile and volumetric mass, a series of "urban villas" with large planting terraces. The whole ensemble of distinct layers, slightly shifted off center, produces a dynamic effect, which, combined with traditional materials and window openings, marks the project as being of Milan today. The S-form of the building was determined through a series of studies seen in diagrams that show how it meets both program and zoning requirements and also incorporates an early 20th century classicist building on the site. This building becomes part of the condominium, as well as its formal entrance. The public garden to the west now "ends" in the curve of the condominium's west elevation.

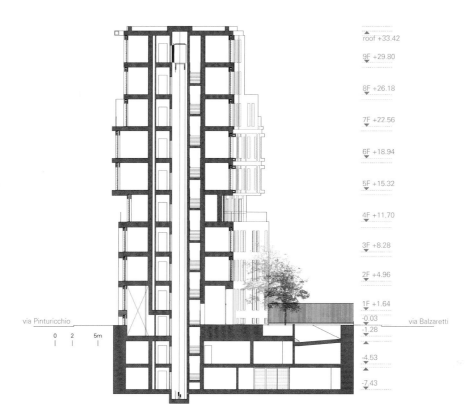

A-A' 剖面图 section A-A'

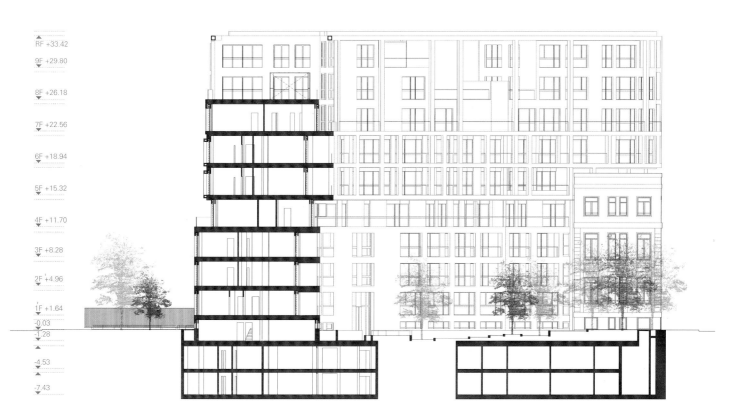

B-B' 剖面图 section B-B'

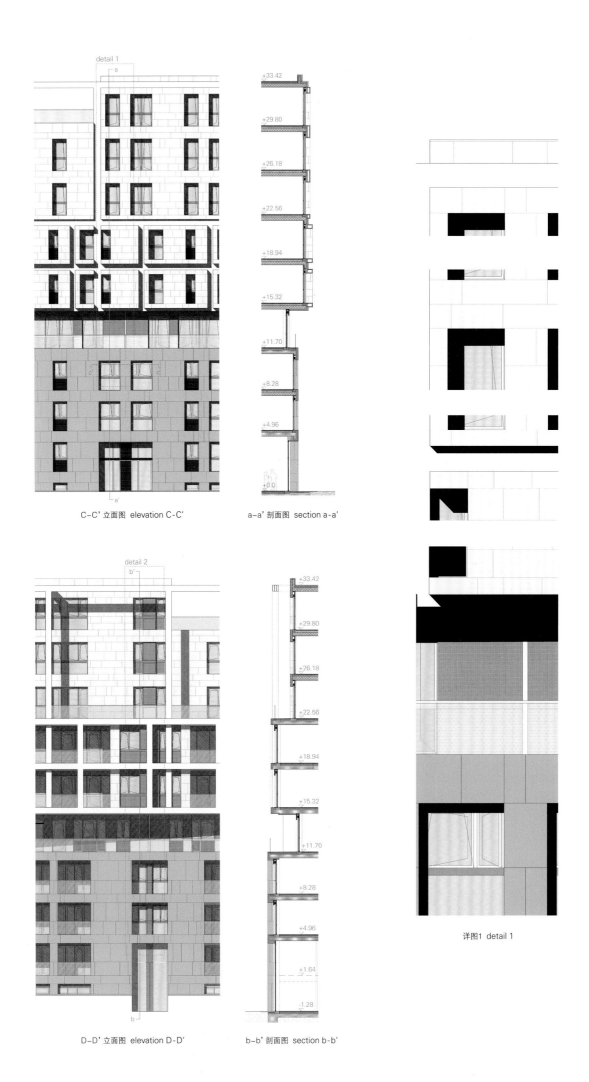

C-C' 立面图 elevation C-C'

a-a' 剖面图 section a-a'

D-D' 立面图 elevation D-D'

b-b' 剖面图 section b-b'

详图1 detail 1

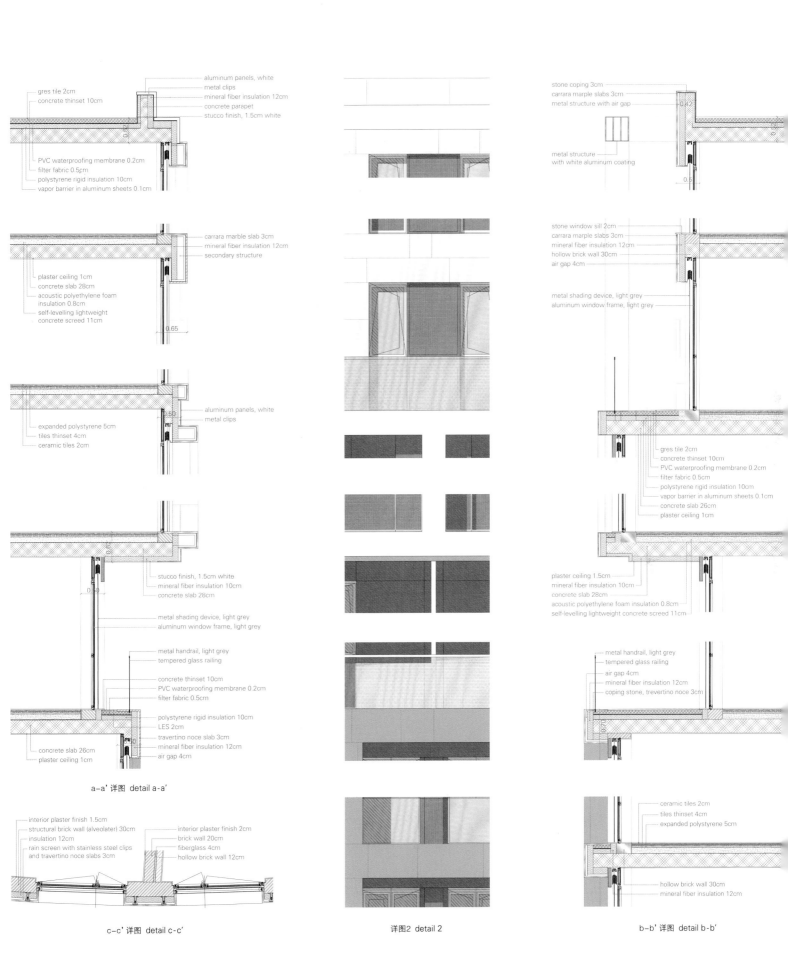

项目名称：Carlo Erba Residence
地点：Milan, Italy
事务所：Eisenman Architects, Degli Esposti Architetti, AZstudio
总承包商：Phase 2 and completion – Italiana Costruzioni; Phase 1 – CLE Cooperativa Lavoratori Edili
结构工程公司：Studio d'Ingegneria Associato Ardolino
机电工程公司：Preliminary phase – Sistema Group Engineering; Execution phase – A.T. Advanced Tecnologies; Execution phase – Studio MGM
现场监理：Lorenzo Degli Esposti, Paolo Lazza, Stefano De Vita
项目监理：Phase 1 – Geom. Giorgio Marocchi – Sherpa Engineering; Phase 2 – Geom. Tristano Barroccu
客户：Pinerba s.r.l., a company of Morelli Group
总面积：14,000m²
总建筑造价：40 million euro
竣工时间：2019.7
摄影师：©Maurizio Montagna (courtesy of the architect) - p.66, p.69, p.72, p.73, p.74, p.78, p.79, p.80~81; ©Marco de Bigontina Esperienza-Drone (courtesy of the architect) - p.67, p.70, p.71

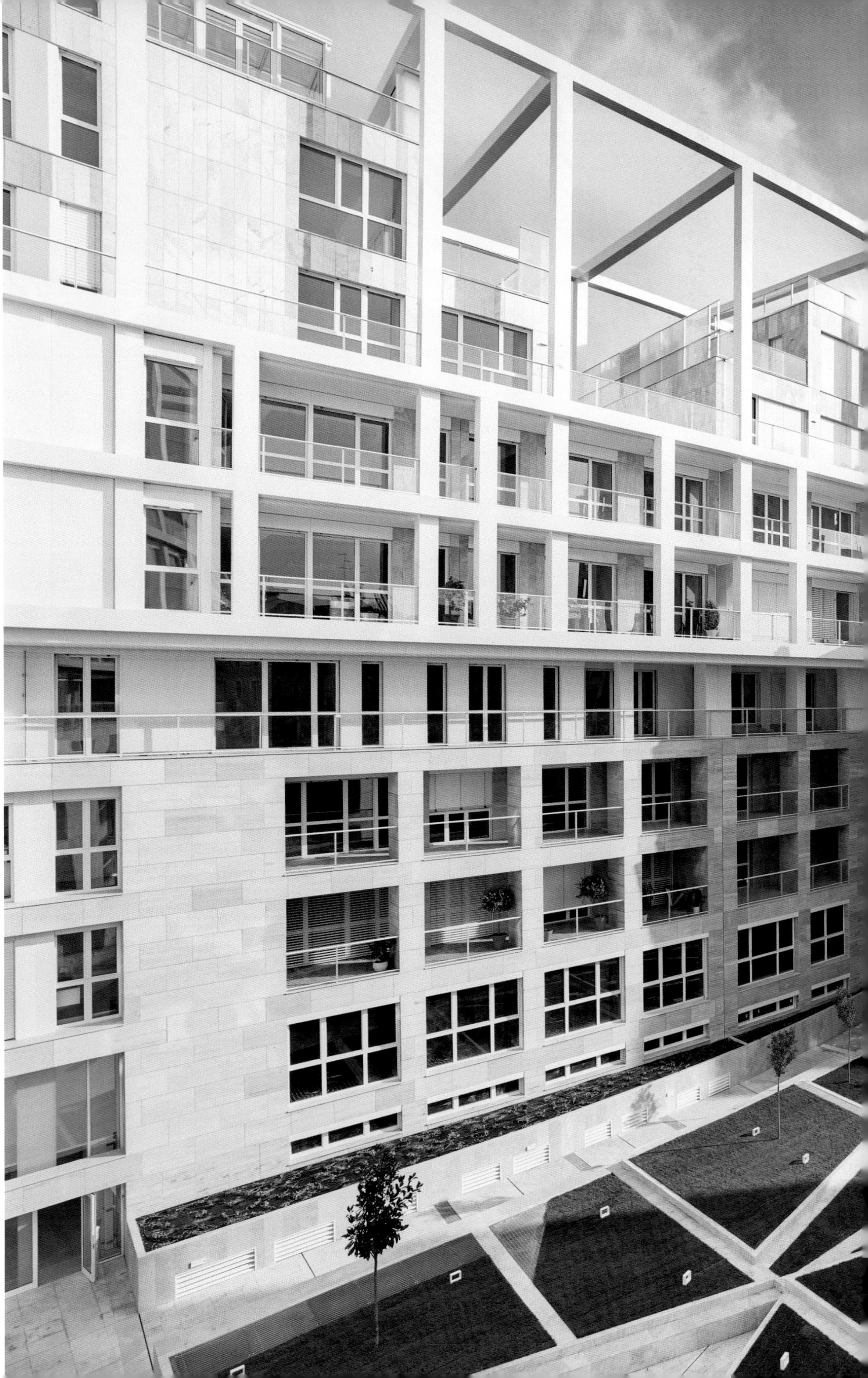

Fenix一期项目
Fenix I
Mei architects and planners

新公寓采用巨大钢结构夹层，坐落于旧日鹿特丹码头仓库之上
New loft apartments in Rotterdam atop a historical quay warehouse interlayered with gigantic steel structure

兴建于Fenix大仓库上的九层公寓经过改造，现已成为混合用途的文化热点，成为鹿特丹水域旁的一个新地标。新建筑大约有45 000m²的混合用途空间，设计大胆，极具标志性，其程度不亚于建于1922年的老建筑。原建筑为服务于荷美航线的旧金山仓库，结构长度曾达360m，令人印象深刻。但由于战火，原建筑遭受了部分损毁，虽历经修复，仍不免在20世纪80年代的港口变迁中遭到废弃。

后来，这座建筑被再次开发，其中一项任务就是搭建一个规模庞大的钢台结构，在上面建造了新的混凝土建筑，并按买方意愿，在建筑中设计了大小不等的公寓间。建筑规模如此之大，设计布局灵活度却如此之高，实属欧洲罕见。混合用途空间包括码头住房、办公室、公共停车场、经济型旅馆、餐饮设施和占地广阔的文化集群。

Fenix一期设计包括三个主要部分。第一部分是原有的旧金山仓库，经过翻新之后可用于混合用途。第二部分是原有仓库和上面新建筑之间的夹层，由巨大的空间框架结构构成，容纳了多个公寓间，旁边即是大型庭院花园。第三部分就是上面新建的封闭式建筑，该建筑采用了灵活的混凝土结构，每个公寓间外都有进深2.5m的户外空间。内部走廊连接着街道层的公共人行道，这条人行道长度为40m，连接着城市与码头。

住户从这条人行道可以到达安全透明的入口，然后由此乘坐玻璃电梯，经过仓库屋顶到达庭院，庭院里光线充足，绿意盎然，通途透明，与船坞工业风格强烈的结构形成了鲜明对比。走廊促进了住户之间的社交互动，形成了包容性的社区。庭院则特别采用了不同色泽的木材，是让人安心静气的绿洲。建筑外侧围绕着玻璃制成的阳台，这就是所谓的"缪斯框架"：重复出现的工业钢铁元素不仅符合鹿特丹港的用途和特点，也为用户带来了学习向上、积极梦想的动力。外立面至少使用了516个缪斯框架。

新旧融合，将建筑与周边环境联系起来，是设计的重要一环。在面向莱茵港和韦尔兰的一侧，住宅楼层与周边建筑形成了很好的搭配。项目还恢复了外立面原有的粗制混凝土、大型装载门和信箱式窗户等特点，凸显了原建筑的用途。

建筑的住宅部分共有214间公寓，公寓层高为4m。钢台的框架结构也是公寓的一部分。已有的Fenix码头经过翻修重建，成为高效的公共停车库。建筑共有5间码头住房，采用自由布局，装有原有的装载门，用户可从码头直接由此进入码头住房。Fenix码头现在还容纳了三个著名的文化机构，它们组成了文化集群。最终的设计效果充满智慧，实现了公共空间和私人空间的共存，人们可以在这里自然相遇。

可持续性是项目秉持的一大基本原则。贯彻落实这一原则的切口在于对原有建筑的最大化重新使用。新建筑能够适应未来的需求变化，而建筑空间则实现了灵活划分。

Fenix一期项目经过优化后，阳光可以照射进庭院和公寓。玻璃外立面由高性能太阳能控制玻璃构成，能够阻挡阳光热量，减少制冷需求。屋顶花园和庭院垂直的绿色立面设计营造了健康舒适、容纳自然的居住环境，同时可以有效过滤空气中的颗粒污染物。绿植屋顶能够收集雨水，以便重复使用。公共区域安装了可以蓄冷蓄热的热回收机械通风系统。

Nine stories of loft apartments placed upon the monumental Fenix warehouse, now transformed into a mixed-use cultural hotspot, forms a new landmark along Rotterdam's waterfront. With almost 45,000m^2 of mixed-use space, the new building is as bold and iconic as the historic building was in 1922. The former San Francisco warehouse, built for the Holland-America Line, was an impressive 360m-long structure, but was partially destroyed by war and fire. Despite restoration, the warehouses fell into disuse in the 1980s with the movement of the port. As part of the redevelopment, an immense steel table structure has been built, on top of which is a new concrete building containing the variously-sized loft apartments, which are developed according to buyers' wishes. This high degree of flexibility is unique in Europe at this scale. The mixed-use space comprises quay houses, offices, a public parking garage, a B&B, catering facilities and an extensive culture cluster.

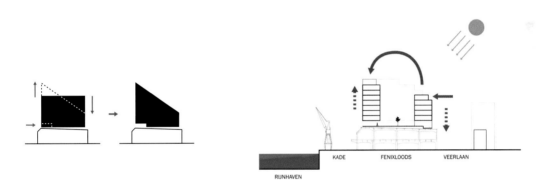

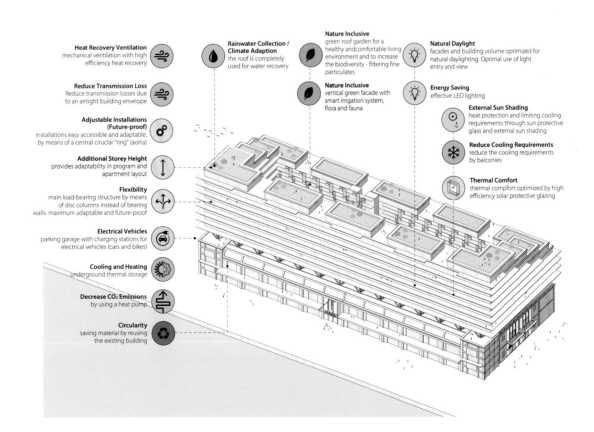

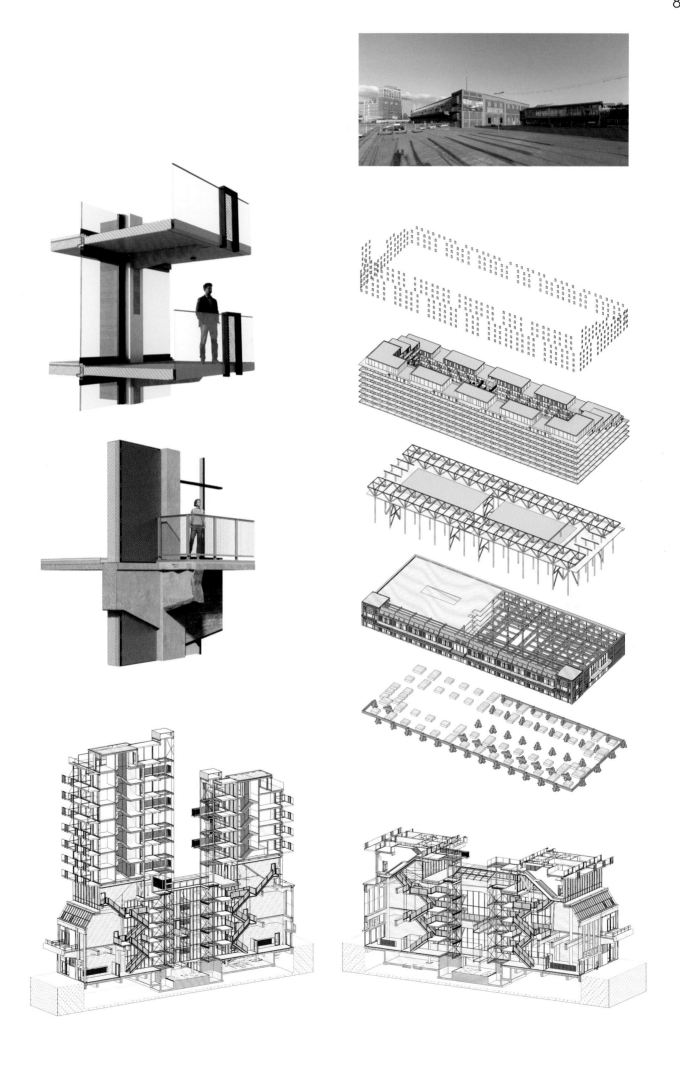

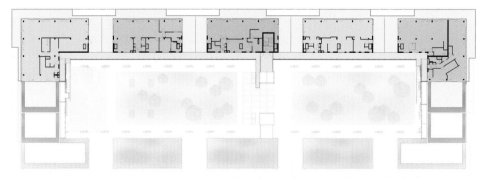

九层 eighth floor

公寓 apartment

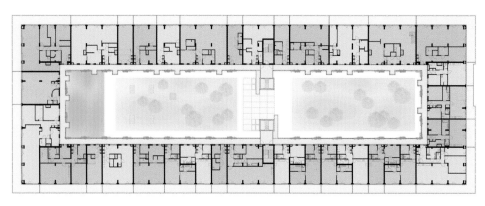

五层 fourth floor

公寓 apartment

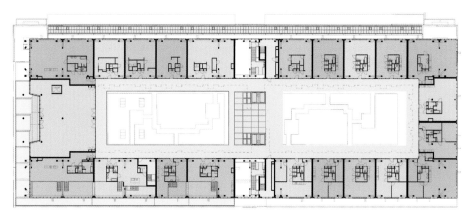

切口层 incision floor

公寓 apartment

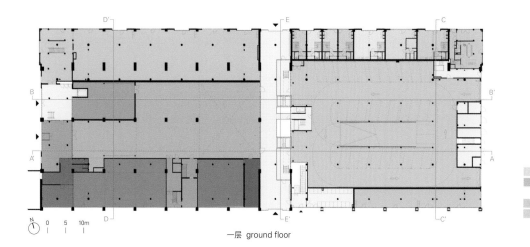

一层 ground floor

住宅 apartment
艺术大学/Rotjeknor马戏团 Codarts/Circus Rotjeknor
商业区 commercial
停车场 parking

公寓概念 loft concept

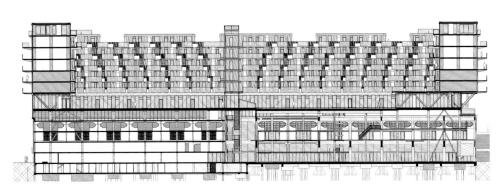

A-A' 剖面图 section A-A'

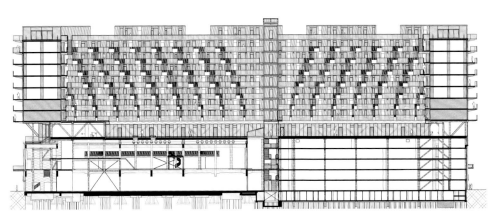

B-B' 剖面图 section B-B'

C-C' 剖面图 section C-C'

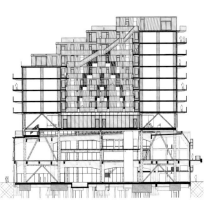

D-D' 剖面图 section D-D'

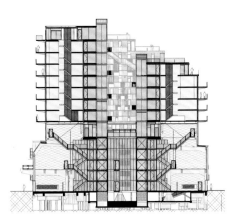

E-E' 剖面图 section E-E'

The design for Fenix I consists of three main parts. Firstly, the existing San Francisco warehouse renovated for mixed use. An interlayer consisting of a gigantic spaceframe structure separates the existing warehouse and new volume above. This accommodates loft apartments adjacent to a large courtyard garden. On top of this spaceframe a new enclosed building block of flexible concrete construction arises, with loft apartments each with 2.5m-deep outdoor spaces around them. The interior gallery connects to a public passageway on street level, a 40m-long connector between the city and the quay. From the passage, the residents reach the secure and transparent entrance to the glass elevators, which lead the residents through the roof of the warehouse into the courtyard. This is characterized by light, greenery and transparency, contrasting with the strong industrial structures of the dockyard. The gallery encourages social interaction between the residents and creates an inclusive community. The courtyard is a calm oasis, emphasized by using different tones of wood. The balconies surrounding the outer side of the building are made of glass and so-called "muse frames": repetitive industrial steel elements that invite users to lean and daydream. and which fit the character and the use of the port. In total no less than 516 muse frames were added to the facade.

Merging old and new was an important part of the design, connecting the building with its surroundings; on the Rijnhaven as well as the Veerlaan side, the residential building

layers were matched to the neighboring buildings. Original characteristics of the facade, such as the brute concrete, large loading doors, and letterbox windows, are restored, emphasizing references to the former building use.

The residential volume comprises 214 lofts, with a height in the apartment of 4m. The frame structure of the steel table construction is unmistakably part of the apartment.

Part of the existing Fenix Docks warehouse was rebuilt to provide an efficient public parking garage. There are five quay houses with a free lay-out, fitted with original loading doors accessible directly from the quay, and Fenix Docks now offers room to three well-known cultural institutions, which together form the Culture Cluster. The result is a smart design with both communal and private spaces, where spontaneous meetings can take place.

Sustainability was a fundamental principle, starting with maximum reuse of the existing building. The new building is future-proof and adaptable; the space can be flexibly partitioned.

Fenix I has been optimized to allow daylight to enter the courtyard and the apartments. The glass facades are made of high-performance solar control glazing, blocking solar heat and reducing the need for cooling. The roof gardens and vertical green courtyard facades stimulate a healthy, comfortable and nature-inclusive environment, while filtering particulates from the air. The green roofs collect rainwater for reuse. In the communal areas, a heat-recovery mechanical ventilation system with heat and cold storage is applied.

项目名称：Fenix I / 地点：Veerlaan/ Rijnhaven, 3072 ZP, Rotterdam, The Netherlands / 事务所：Mei architects and planners / 设计团队：Robert Winkel, Robert Platje, Michiel van Loon, Ruben Aalbersberg, Sean Bos, King Chaichana, Ed de Rooij, Kasia Ephraim, Danijel Gavranovic, Jan Hoogervorst, Rutger Kuipers, Arjan Kunst, Riemer Postma, Rob Reintjes, Adriaan Smidt, Daam van der Leij, Ruben van der Plas, Lore van de Venne, Menno van der Woude, Reinoud van der Zijde, Johan van Es, Roy Wijte / 总承包商：Heijmans Woningbouw

顾问：structural engineer – ABT Delft; MEP – Techniplan; building physics – LBP|Sight; historical research – Suzanne Fischer / 供应商：CSM Steel structures, reynaers aluminium façade systems, HCI prefab concrete, Trahecon fences / 客户：Heijmans Vastgoed / 总楼面面积：40.500m² / 造价：€55.000.000 / 竣工时间：2019.5
摄影师：©Ossip van Duivenbode (courtesy of the architect) - p.84~85; p.88, p.91 middle, bottom, p.94, p.95, p.96~97 right-top; ©Marc Goodwin (Archmospheres) (courtesy of the architect) - p.83, p.86, p.87, p.90, p.91 top, p.96~97 left, right-bottom

Bottière Chénaie住宅综合体
Bottière Chénaie Housing Complex

KAAN Architecten

南特东北部如巨石般的混凝土住宅楼，表面设计极具凝聚力
Monolithic concrete residential buildings accompanied by a cohesive repertoire of surfaces in north-eastern Nantes

项目周边基建环绕，有轨电车、老旧铁路线和圣露西主干道形成了隔绝带，孤立了项目地块。为此，KAAN建筑师事务所提出了平衡建筑体量构成、聚合混合用途项目和公共空间的设计，在一片主要为住宅的区域内，使之与库斯托指挥官广场一起，成为南特的第二个焦点。

项目布局极其紧凑，采用了多功能基座，基座内设有商业空间、超市和停车场。基座的上面是两栋五层的住宅楼，其中共有172间公寓。第一栋是面向市场出售的住房，第二栋为公益住房综合楼。

项目西北侧较大的一座住宅楼围绕宽敞的庭院铺开。庭院内种有树木，外周设有直通两座住宅楼的通道，在通道上可以俯瞰周围绿色的环境。该环形解决方案的灵感来自于荷兰住宅建筑传统，不仅做到了因地制宜，而且极大地减少了垂直连接，从而让住宅单元可以实现多元化的配置；无论是私人户外空间还是带有生活区的凉廊，这里一应俱全。建筑师在方向布局上采用了异于常规的设计，而且还设计了天井，自然日光可以通过天井洒向底层楼层。这种垂直方向的断裂设计创造了非正式空间，大大促进了居民的社交互动。项目东侧的第二座住宅楼则有39间公益住房公寓，这些公寓根据位置的不同，都能从双朝向设计中受益，这是一种不同的凉廊建筑类型，住户可以俯瞰芳草绿树，或远眺南特城。最后是位于西南角的办公楼，建筑围绕中央核心筒组织，增强了建筑体量并置效果，补全了全新的城市景观。

Bottiére Chénaie看似简单的建筑体，实则采用了极具凝聚力的表面设计，形成的统一效果堪称经典。虽然项目第一眼看上去体积庞大、平淡无奇，但设计方采用了优质的外立面设计，成功丰富了项目内涵。外立面由规则的60cm×60cm预制混凝土梁柱结构构成，形成了大而透明的表面，以及独特的、完全玻璃质地的转角。项目布局极其紧凑，整个建造流程都实现了经济标准和环保建议的最优化效果。外立面为建筑室内提供了充足的阳光照射，建筑全天候与大自然连接。

面向庭院和通道的垂直表面覆盖了灰色木材，凸显了浓郁的家庭氛围，不仅提高了室内和室外的双重性，同时也加强了内部公共空间的作用，使之成为整个项目画龙点睛之笔。

作为建筑设计方，KAAN建筑师事务所精心选择材质，平衡色泽格

调，采用功能设计，注重项目细节，以雅致优美的手法，塑造了项目全新的社区身份，使之成为魅力十足的中心，并无缝融入南特不断发展的郊区中。

Encircled by existing infrastructures such as a tramway, an old railway line and the major artery of Route de Sainte Luce, the site initially appears isolated from its surroundings. By proposing a balanced composition of volumes and bringing together mixed-use programs and public spaces, the architects have created a second urban focal point, along with the Place du Commandant Cousteau, in a mainly residential neighborhood. Extremely compact, the project features a multifunctional plinth, composed of commercial spaces, a supermarket and a car park. Above this base rise two five-story residential blocks comprising 172 apartments in total. The first volume is dedicated to market-rate housing, while the second hosts a social housing complex.

To the north-west, the larger housing volume unfolds around a spacious courtyard planted with trees, where external walkways overlooking the green surroundings provide access to the dual aspect apartments. Inspired by the Dutch residential

项目名称：Bottière Chénaie Housing Complex / 地点：190 Route de Sainte Luce, 44300 Nantes, France / 事务所：KAAN Architecten – Kees Kaan, Vincent Panhuysen, Dikkie Scipio / 项目团队：Dante Borgo, Sebastiaan Buitenhuis, Marc Coma, Sebastian van Damme, Paolo Faleschini, Marylène Gallon, Renata Gilio, Narine Gyulkhasyan, Sophie Ize, Jan Teunis ten Kate, Wouter Langeveld, Julie Le Baud, Yinghao Lin, Aimee Mackenzie, Elsa Marchal, Ismael Planelles Naya, Ana Rivero Esteban, Cécile Sanchez, Yannick Signani, Christian Sluijmer, Joeri Spijkers / 当地建筑师：Atelier Forny / 城市规划：Pranlas-Descours Architect & Associates, Atelier Bruel Delmar / 总承包商：Angevin Entreprise Générale, Rennes / 施工监理、财务顾问、结构工程公司：AIA Ingenierie / 消防控制：Socotec / 技术设施顾问、声学设计：SerdB – Groupe Gamba, Saint Sébastien-sur-Loire / 客户：OCDL – Groupe Giboire / 功能：172 apartments (133 market-rate housing + 39 social housing), offices, commercial spaces, public squares and car park / 造价：EUR 22,000,000 / 总楼面面积：17,200m² / 竞赛时间：2012~2013 / 设计时间：2014.7—2016.10 / 竣工时间：2019.12 / 摄影师：©Sebastian van Damme (courtesy of the architect)

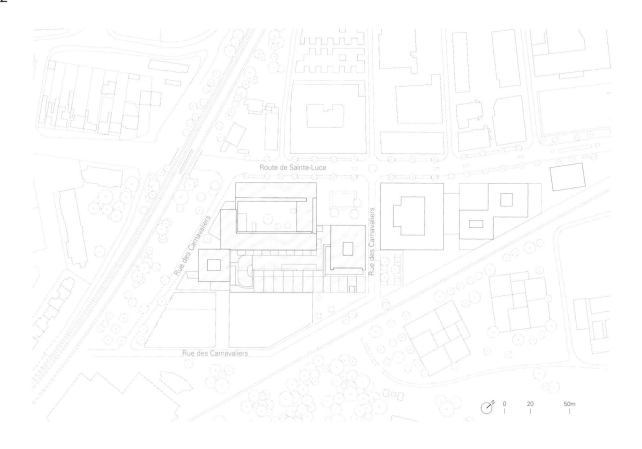

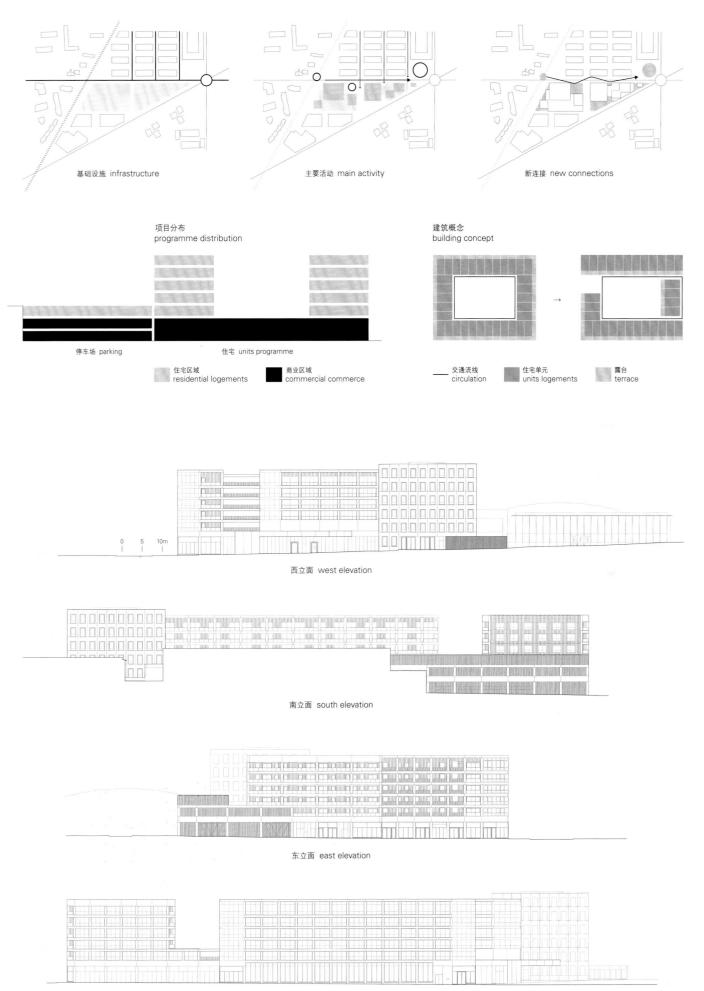

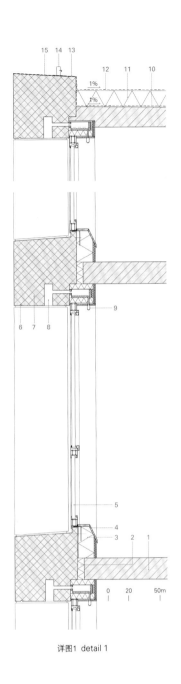

详图1 detail 1

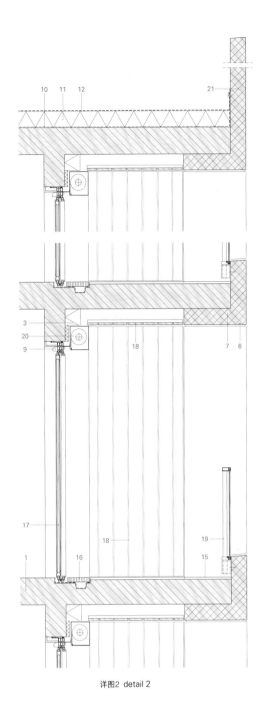

详图2 detail 2

1. in-situ concrete slab
2. thermal bridge breaker
3. insulation - mineral wool
4. window sill - plywood plate
5. aluminum window - fixed and concealed opening frame, anodized finishing
6. water drain profile
7. beam - prefabricated concrete element, waterproof finishing
8. air joint
9. window integrated ventilation system
10. vapor barrier
11. insulation - polyurethane foam
12. self-protective double-layer waterproofing system
13. flashing
14. fixed support for a safety net
15. liquid sealant
16. gutter - liquid sealant
17. aluminum sliding door - anodized finishing
18. wooden cladding - grey tinted douglas fir wood
19. balustrade - aluminum anodized structure and glass panels
20. roller shutter casing
21. parapet - prefabricated concrete element, waterproof finishing
22. column - prefabricated concrete element, waterproof finishing
23. in-situ concrete wall
24. aluminum profile

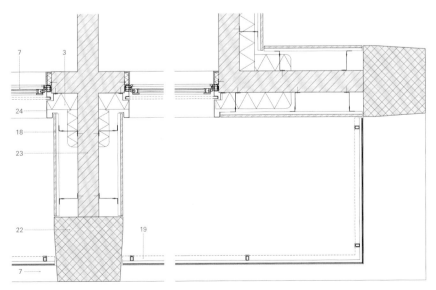

详图3 detail 3

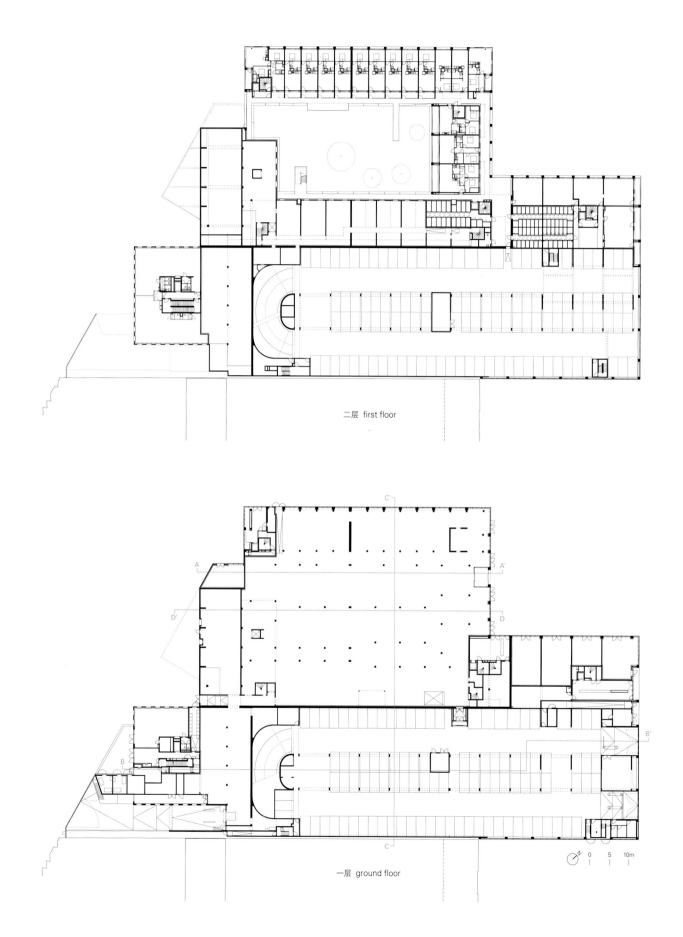

二层 first floor

一层 ground floor

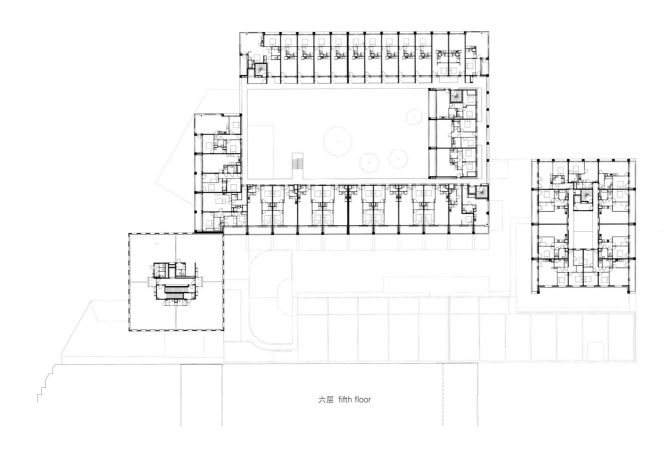

六层 fifth floor

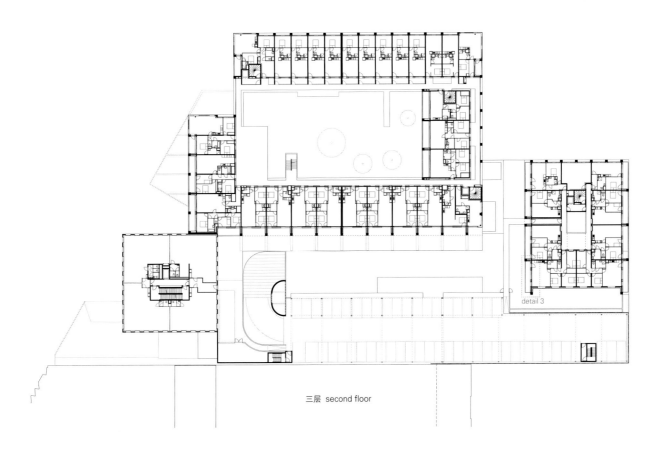

三层 second floor

architecture tradition, this circulation solution is adapted to its context and minimizes vertical connections, allowing diverse configurations to the housing units – from private outdoor spaces to loggias with living areas. Moreover, these anomalies in orientation are flanked by light wells spreading natural daylight to the lower floors: a vertical rupture that generates informal spaces to foster residents' socialization.

To the east, the second volume accommodates 39 social housing apartments, which, according to their position, benefit from a double orientation, a different typology of loggias and look out over the trees or towards the city. Lastly, to the south-west, the office block, organized around a central core, reinforces the volume juxtaposition and completes the new urban landscape.

Monumental in its consistency, the simple massing of Bottière Chénaie is accompanied by a cohesive repertoire of surfaces. Monolithic at first glance, the project is enriched by high quality facades composed of a regular 60cm x 60cm post-and-beam structure in precast concrete that generates large transparent surfaces and singular fully glazed corners. The project's compactness allowed the optimization of economic standards and environmental recommendations throughout the whole building process. The facades provide abundant daylight to the interiors and naturally link them to the city, both during the day and at night.

The vertical surfaces which face the courtyards and walkways are clad in grey stained timber giving them a characteristic domestic feel and reinforcing the duality of exterior and interior while reinforcing the inner public space as the beating heart of the project.

Thanks to a meticulous choice of materials, a balanced color tuning, functional design and attention to details, KAAN Architecten has shaped the new neighborhood identity with an elegant intervention, transforming it into an attractive center that blends seamlessly into Nantes' growing suburbs.

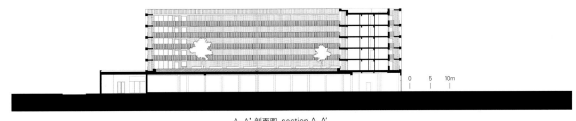

A-A' 剖面图 section A-A'

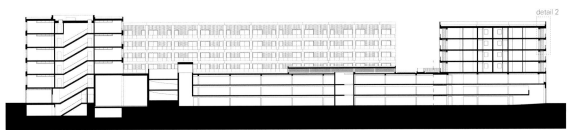

B-B' 剖面图 section B-B'

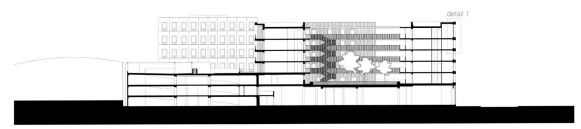

C-C' 剖面图 section C-C'

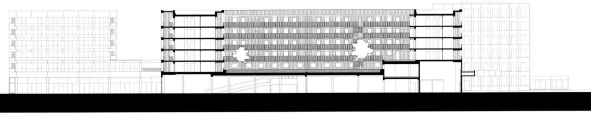

D-D' 剖面图 section D-D'

Denizen Bushwick公寓楼
Denizen Bushwick
ODA

参与社区互动，联结社区居民——布鲁克林酿酒厂旧址上新建的住宅综合楼
A new residential complex designed to engage the community on the former brewery site in Brooklyn

Denizen Bushwick项目位于布鲁克林历史悠久的Rheingold酿酒厂，共有93 000m²的公寓单元，其中20%为保障性住房。项目容纳了大量向社区开放的公共空间，公园将项目一分为二，形成了两楼夹一绿道的布局。设计方还添加了一系列蜿蜒曲折、互连互通的庭院，点缀了项目建筑群，这些庭院最终都通向绿色通道。位于纽约城中的这两栋大楼整体却采用了典型旧城区核心中的交织型街道布局。

庭院景观郁郁葱葱，部分有顶的通道和走廊穿行在树荫之下，形成了一个又一个广场，其中设置的便民设施将在这一片愈发活跃的区域中激发居民的社区感。按照主导设计原则，项目在采用典型的网格式布局的同时，鼓励休闲、促进发现，从而进一步完善了结构和效能。

项目的目标是：打造高度孔隙化的建筑布局，为社区提供活动和互动的平台，成为社区密不可分的一部分。在修建幽蜒互连的庭院时，设计方用一条绿色通道将项目一分为二，促进了公共活动的开展，打造了名副其实的城中之城。

ODA事务所的社区公共事务机构OPEN一贯热心支持以下四类项目：社区绿色空间、媒体与技术、设计与建筑、公共艺术。此次，OPEN也积极参与了该综合楼项目15幅大型壁画的设计。15幅壁画贯穿了整个综合楼，装饰了高达七层的走廊墙面、一个天花板和公园附近的一处停车场。添加壁画元素的目的是为整个区域带来明媚色彩，带来勃勃生机，因此壁画规模很大，人们在庭院中即可看见。除此之外，人们还可以从公园中看到5幅壁画，这正是设计方为活跃社区气氛、鼓励当地社区互动参与而做出的策略性决定。

在社区建设与社区参与中，便民设施也起到了重要的作用。项目设有一处咖啡店、一面攀岩墙、一个拳击台、一家微型酿酒厂、多个游戏室等设施。项目屋顶面积为6500m²，设有迷你高尔夫球场、露天烤肉吧、无土栽培城市农场和数量众多的座位。

一踏入Denizen Bushwick公寓楼大门，新的空间就呈现在眼前了。它的门厅极有社区感，采用了丰富的当地艺术元素，整座建筑创意十足，公园景区层层叠叠，零售区和娱乐区多种多样。项目不仅包含保障性住房，还有着优越的地理位置，不仅是创意人才和职场新秀的绝佳选择，而且在繁华喧闹的都市中打造了强烈的社区归属感；它不仅能促进社区参与，还为居民提供了开阔的视野。Denizen Bushiwick住宅楼是优美与功能并重的经典之作。

Situated on the former site of Brooklyn's historic Rheingold Brewery, Denizen Bushwick will generate about 93,000m² of apartment units in the area, 20% of which are classed as affordable housing. The project will host a multitude of communal spaces open to the neighborhood, while a public park will bisect the development creating a green promenade and two blocks. These masses are further perforated by a sequence of meandering, interconnected courtyards which ultimately lead to the promenade. Over the pair of these New York city blocks, the architects have superimposed the layout of woven streets in a typical old town core.

Within the courtyard areas, lushly landscaped and partially covered walkways and corridors will give way to a parade of plazas, and accessible amenities designed to continue to promote a sense of community in this increasingly vibrant area. Complementing the structure and efficiency of a more typical grid, the layout will encourage both leisure and discovery, the guiding principles of the design.

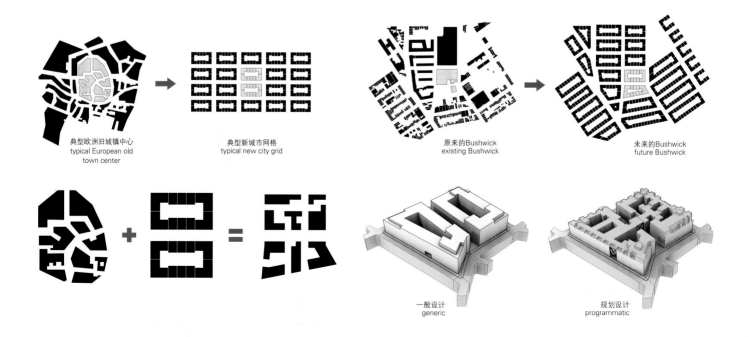

典型欧洲旧城镇中心 / typical European old town center

典型新城市网格 / typical new city grid

原来的Bushwick / existing Bushwick

未来的Bushwick / future Bushwick

一般设计 / generic

规划设计 / programmatic

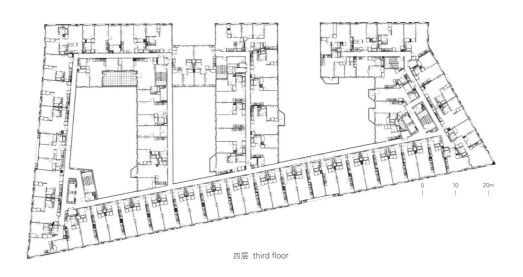

四层 third floor

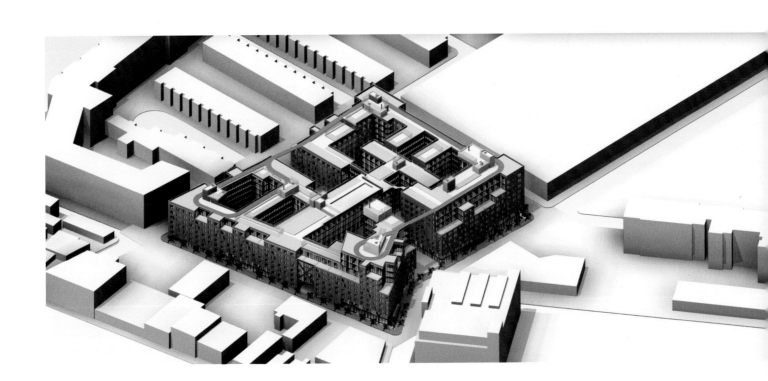

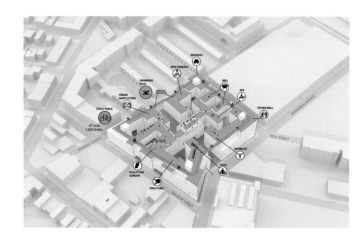
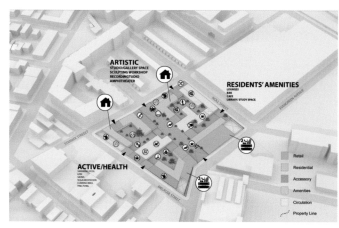

Denizen aims to become an integral part of the neighborhood by creating a highly porous building where the community can find a platform for activity and interaction. With the implementation of meandering, interconnected courtyards, a bisecting green promenade, and communal activities, it will be a veritable city within the city.

Heavily involved with the design of the 15 large-scale murals in the complex is the nonprofit organization, ODA's Public Engagement in Neighborhoods[OPEN]. OPEN enthusiastically supports ventures in the following four categories: community green space, media and technology, design and architecture, and public art. The murals are strewn throughout the complex, adorning seven-story corridor walls, one ceiling and a parking ramp near a public garden. With the intent of bringing vibrant color and character to the area, all of the murals are on a massive scale and can be viewed from the development's multiple courtyards. In addition, five of the murals are visible from parks that are open to the public, a strategic design choice meant to enliven and engage the local community.

Amenities also play a large part in the design for community building and engagement. Denizen has a coffee shop, game rooms, a rock climbing wall, a boxing ring, a fully functioning microbrewery and more. The rooftop boasts 6,500m² of space, including a mini-golf course, barbeques, a hydroponic urban farm and plenty of seating.

From the moment of entry, Denizen Bushwick provides a new kind of space; its lobby creates a sense of community, rich with local art, innovative architecture and layered vistas of the parks, retail and recreational spaces. Joining together affordable housing with a prime location for creative people and budding professional people alike, Denizen creates a sense of community within a bustling city. Designed to engage the community and offer expansive views for residents, Denizen Bushwick is as beautiful as it is functional.

项目名称：Denizen Bushwick
地点：54 Noll Street, Brooklyn, NY, USA
建筑师：ODA – Eran Chen
项目团队：Ryoko Okada, Yaarit Sharoni, Kristina Kesler, Hadas Brayer, June Kim, Brona Waldron, Chris Berino, Jennifer Endozo, Carolina Moscoso, Roman Falcon, Sojin Park, Tulika Lokapur, Chia-Min Wang, Seung Bum Ma, Charles Burke, Adrienne Milner, Emma Pfeiffer, Alex Sarria, Jaehong Chung, Dawoon Jung, Joshua Wujek, Asuncion Tapia, Heidi Theunissen, François Blehaut, Brian Lee, Joohwan Seo, Paul Kim, Steven Kocher, Vi Nguyen, Shraddha Balasubramaniam, Gökçe Saygın Batista
结构工程师：McNamara Salvia
承包商：NYEG Corp. & All Year Management
客户：All Year Management
用途：residential
总楼面面积：111,484m²
设计时间：2015—2017
施工时间：2017—2019
摄影师：©Eric Laignel (courtesy of the architect) - p.121, p.124~125, p.126~127, p.128, p.129, p.130~131; ©Miguel de Guzman (courtesy of the architect) - p.118~119, p.122, p.130, p.131

幼儿园：灵活、安全

Kinderg
Being Fl
Being S

在当代生活中，幼儿园和学校是幼童和少儿日常生活中重要的活动场所。他们和日托人员在那里度过的时光应该是高质量的；孩子们在那里受到保护，也有行动的自由，父母相信他们把孩子送到幼儿园和学校很安心。除家庭之外，孩子们还将在这里认识这个世界，迈出探索和理解这个世界的第一步，并开始他们生命中一个重要的阶段：社交关系。

幼儿园的建筑和设计对孩子们的学习很重要，因此也在孩子们的成长过程中起到了至关重要的作用。如何才能设计出既安全，又能激发好奇心、鼓励游戏、整合最新的技术、培养特定的

In contemporary life, kindergartens and schools are spaces where toddlers and young children stay for an important amount of their daily life. The time they spend there, together with daycare staff, should be quality time, where they are protected but also have freedom in their movements and where parents trust they can safely leave their kids. Here, outside of the family nucleus, they will awaken to the world, taking their first steps to explore and understand it and start an important phase of their lives: social relationships. Architecture and design are the great hosts of the learning process and therefore with a fundamental role in children's development. How can the planned spaces offer security,

Würenlingen幼儿园_Kindergarten Würenlingen/Malte Kloes Architekten + Estrada Reichen Architekten
Storey's Field中心与爱丁顿托儿所_Storey's Field Center and Eddington Nursery/MUMA
O6A地块幼儿园_O6A Lot Kindergarten/sam Architecture + Querkraft
MUKU托儿所_MUKU Nursery/Tezuka Architects

幼儿园：灵活、安全_Kindergarten: Being Flexible, Being Safe/Paula Melâneo

技能、激发社交和激发语言的空间呢？要满足所有这些要求，同时实现健康学习环境的诸多其他要求，这对建筑师们来说是挑战。

在快速变化的时代，灵活性是设计构思的一个关键特征。设计的结构应该能够完全满足当地社区的不同背景和需求，并遵循新的教育标准做出相应的改革或者调整。只有这样，建成的基础设施才有资格成为社区得力的辅助性设施。

inspire curiosity, encourage play, integrate updated technology, foster specific skills, motivate socialization and stimulate language? The answer to these questions, among many other requests for achieving a healthy learning environment, is the exercise proposed to the architects.
In fast-changing times, flexibility is a key characteristic in the design response. The designed structure should be able to answer directly to local communities' different contexts and needs, and to follow transformations or modifications required by new education standards.
Only then, will the built infrastructure be qualified to act as a community subsidiary.

灵活、安全
Being Flexible, Being Safe

Paula Melâneo

"教育设施既要适应今天已知的和那些已经得到认同的需求,也要着眼未来不确定的需求。"
—— 经合组织,《设计学习空间: 55个典范式教育设施》(2001年)

"建筑物应该为教育和学习提供一个总体建筑架构,同时还能够灵活地适应不断变化的需求,甚至在空间感上主张追求畅通的通道而不是设计任务书中所要求的那些狭小的空间。"
—— 赫尔曼·赫茨伯格,《赫尔曼·赫茨伯格的学校》(2009年)

数年来,儿童教育模式一直在改变。从一个等级森严、有条不紊和纪律严明的体系(其中学校是公共代表的典范),而演变成了一种更加非正式、更感性的以及参与式的体系,在这样的体系中,空间更加人性化,能提供更加广泛的体验内容。近几十年来,随着城市空间的增长和新技术的开发和应用,社会奔腾发展。学习空间如何设计才能适应这些变化呢?

单一功能的空间设计很难(同时也比较昂贵)适应迅疾变化的新需求。而且,单一的学习空间设计结构无法提供更丰富的使用机会,诸如用于社会交流、玩耍、跳舞或者专门用于某一项特定的任务,如阅读、听故事,甚至举行一些综合性的活动。专注于空间灵活性的设计理念才更有可能成功。但是,长期从事教育空间研究的赫尔曼·赫茨伯格认为,这种灵活性"需要稳定的设计框架,来适应这种变化"。[1]设计师们要想解决这个问题,首先要研究当地的规则和那些空间的具体标准,还要考虑有多种可能的空间的设计: 如何运用轻型材料和结构或者是模块化家具来进行空间分割,创造出不同的布局类型; 在哪里光可以自由进入或者更容易控制; 在哪里声音能被反射或者减弱; 在哪里可以对"中间"空间(例如,交通流线或者走廊)进行栩栩如生的设计,并将其作为辅助性的非正式学习区域。这些多功能空间更容易满足那些意想不到的、紧急的变化需求,对使用者来说既高效又舒适。

"Educational facilities need to accommodate both the known and identifiable needs of today, and the uncertain demands of the future." – OECD, Designs for Learning: 55 Exemplary Educational Facilities (2001)

"The building should provide a general framework for education and learning, while being flexible enough to respond to changing demands and even in a spatial sense hold out a suggestion of pursuing avenues other than those laid down in the brief." – Herman Hertzberger, The Schools of Herman Hertzberger (2009)

Over the years, children's education has been shifting its paradigms, from a hierarchic, orderly and strongly disciplined system, where schools were models of public representation, to a more informal, sensorial-based and participatory one, where spaces are humanized and offer a wide range of experiences. Over recent decades, changes are still speeding up as urban spaces grow and as new technologies are being developed and adopted by societies. How can the learning spaces be projected to go along with those changes?
Mono-functional spaces are harder (and expensive) to adapt to new demands that, sometimes, are happening too fast. Also, they don't offer as many opportunities for learning as a space where you can, at turn, be socializing, playing, dancing or concentrating on a specific task, such as reading, listening to a story or even combining different activities. Concepts more focused on space flexibility are most likely to be successful. However, Herman Hertzberger, who has long studied the educational spaces, advocates this flexibility creates the "need for a stable framework to capture that changeability"[1]. Starting with dealing with local rules and specific standards for those spaces, designers try to solve that issue by thinking about spaces that are prepared for multi-possibilities: where the partitions to achieve different types of configuration can be done with light structures and materials or modular furniture; where light can freely enter or easily be controlled; where sound can be reflected or muffled; where "in-between" spaces, such as circulations or hallways, can be animated and used as complementary and informal learning areas. These multi-functional spaces are more liable to satisfy non-predicted and urgent changes and still be highly efficient and comfortable to users.
Nowadays, education can be seen as a continuous process that happens beyond the school. This induces the school's surroundings and the city itself to be part of that educational process: "not only does the school become a small city but the city becomes an exceedingly large school"[2]. This idea accords an important mission to the immediate surroundings of the school in children's learning context. Such as some of the projects featured in this

如今，教育可以被看作是发生在校内校外的一个连续过程。这促使学校的环境和城市环境本身融为一体，成为教育过程的一部分："不仅学校成为一个小城市，与此同时，城市亦变成了一所超大的学校"。[2] 这一理念在孩子们的学习环境中为学校的周边环境赋予了重要的使命。例如，这一部分中展示的一些项目，教育设施可以与同一综合建筑中的其他功能相结合。这种共生关系颇有裨益，但也必须确保安全，让孩子们感到舒适、安全、自在。尽管建筑单独服务于不同的用户，但如果当地社区环境对孩子们来说非常安全，或者孩子们的领地可以封闭起来与外界隔离开，并使内部空间成为孩子们的庇护所或保护中心，那么该建筑就可以作为一个和谐的聚集场所使用。

在房间中和在有特点（例如，有隔墙、通透）的交通流线空间中进行有创意的设计可以让（或者阻止）大家进行眼神交流和社交聚会。这可以让孩子们或者成年人之间畅通地交流。有创意的设计可以将户外空间和大自然引入室内，比如，自然景观或者相连的花园、日光或者声音。此外，空间尺度的大小也会影响孩子们的安全感，例如，比较熟悉的家居大小的空间会使孩子们感到更温暖、更舒适、更轻松，而设计一个大比例房间则要考虑到环境要更加安静，设计也要更加有意识。

如今，安全问题比以往任何时候都更加受关注。无论是从身心方面，还是从最新的健康角度来看，安全都成为全球关注的焦点。2020年的全球疫情向社区提出了新的挑战，特别是在城市环境中。建筑设计师和决策者们面临的是：要使公共空间的设计能够迅速适应新的生活规则，需要重新考虑特定的小型社区的安全问题，比如，幼儿园和学校。社交距离和社交活动需要新的意义和形式，最灵活的建筑可以具备最充分的使用条件，迅速满足最新的变化需求，而且不会影响到孩子们的社交活动和教育发展情况。宽敞的空间、活动自如的交通流线空间、开放的场地以及通风良好的区域，这些都使孩子们拥有了更多的选择进行他们彼此之间以及与周围环境之间的互动，同时，也有益于孩子们与学校工作人员和他们家庭成员之间的和谐互动。

令人兴奋的学习空间必须为孩子们提供舒适安全的日常活动日程，有进行各种社交活动的机会。在这样的环境中，环境和氛围的创新变换会带来不同的新体验。使用透光材料或窗帘可以创建出一种空灵缥缈的氛围；而不透明的材料会阻挡光线，给人一种密闭的感觉；精心布置砖、瓷砖或者木材等材料可以创造出各种图案以及能给人带来愉悦触感的纹理；运用色彩可以产生不同的图像和视觉效果；创建形象的几何图形可以激发儿童的想象力等等。建筑的特殊设计成为儿童学习过程中的激发元素，能够激发意识、产生熟悉的认知关系、促进新知识的获取、增加感知力以及鼓励孩子们的创造力。

剑桥社区中心与爱丁顿托儿所于2018年由总部位于伦敦的MUMA建筑师事务所设计建成（150页），位于爱丁顿，处在剑桥西北的一个开发区内。近几年，这一地区进行了城市化升级改造，以缓解剑桥的城市发展压力。该项目集两种类型的建筑规划

part, learning facilities can be combined with other functions in the same complex. This symbiosis can be beneficial, but safety must also be ensured and children should feel free and safe at the same time. The structure can act as a harmonious gathering space, although serving different users separately, if the local community is part of the safety net for children, or children's territory can be set apart and closed to the outside, opening the inside as a shelter or a protective nucleus for youngsters.

Working creatively in rooms and circulations with features such as partitions and transparences can enable (or disable) eye-contact and social meeting. This allows for the establishment of communication with other children or adults. It can bring the outer space and nature inside, such as the landscape or a contiguous garden, daylight or sounds. Also, the space scale can influence the sense of security: as for instance a more familiar and domestic scale makes children feel cozier, more comfortable and relaxed, or an imposing large proportion demands more quietness and awareness.

Today, more than ever, safety is a main concern in the global conversation, not only from the psychological and physical perspective but also from the up-to-date health point of view. The year 2020 brought a pandemic health issue to the world that imposed some new challenges to communities, especially in urban contexts. Architects and decision makers were faced with adapting architecture to new living rules for common spaces in a very short time, rethinking safety parameters for specific small communities, such as kindergartens and schools. Proxemics and socializing acquire new meanings and forms, where the most flexible buildings can be the most well prepared to respond quickly to new priorities, without compromising children's social and educational development. Wide surfaces, free circulation spaces, open spaces and well-ventilated areas offer more options on how children interact between themselves and with the surrounding environment, and how they relate to the school staff and with their families.

A stimulating learning space must provide children a comfortable and safe routine, with various socialization opportunities. It is where new experiences are suggested by the environment and atmosphere created. Using light transparent materials or curtains will suggest a more ethereal atmosphere; opaque materials cut out the light and give the feeling of enclosure; designing carefully the placement of the materials – such as bricks, tiles or wood – can draw patterns and create textures pleasant to the touch; applying colors may create pictorial and visual effects; giving shape to friendly geometries can stimulate children's imagination, etc. The specificities of the building are triggers to the learning process; they activate the senses, create familiar recognition relations, motivate new knowledge acquisition, increase perception and encourage creativity in children.

模式于一体，一种是作为社区中心的公共基础设施，另一种是可以容纳100个孩子的托儿所，二者之间明确划分了各自的功能和入口。该项目平面图呈四方形，围绕中央"景观庭院"进行布局，这个庭院是一个大的回廊。该建筑的三个侧翼都供托儿所项目使用，还有一条通道位于中央庭院。正如建筑师所命名的那样，在这里，操场就是一个可以"玩耍的花园"。这里的长椅形状各异，非常有趣，绿色植被尽收眼底。供托儿所工作人员使用的设施与儿童区明显分开。中央"景观庭院"内四周的通道连接着托儿所区域，上面有顶棚可以挡风遮雨，这里是一个过渡空间，孩子们在这里可以进行各种活动。孩子们的教室非常宽敞，被设计成家庭房的形状，让孩子们感到熟悉而亲切。教室里光线充足，有足够自由的活动空间。站在宽大的窗户前，外面青翠的花园映入眼帘。透过卧室内一个个小小的圆窗，孩子们可以一览窗外的世界，不经意间培养了孩子们的好奇心。孩子们的各种感觉也受到激发：一些窗户使用浓烈的颜色描绘出了熟悉的图形；砖、石头、木材和其他材料提供了不同的纹理，供孩子们体验和触摸；教室内的声音由带有图案的天花板吸收；花园中植物的清新气味似乎弥漫了室内空间。

该建筑的另一翼是社区中心，朝向建筑的外侧。在其内部，建筑师们设计了一个封闭的"私密"庭院。这里有绿色的植被，但又与供孩子们玩耍的操场分离开来，互不干扰。这个项目的设计十分理性，同时在几何结构上清晰简洁，既富有趣味，又令人感到愉悦。

如何才能将两个听起来对立的项目结合在一座建筑中呢，比如，一个幼儿园和一个地下停车场？这是一场设计竞赛提出的任务，两个总部位于苏黎世的事务所，即Malte Kloes Architekten建筑事务所和Estrada Reichen Architekten建筑事务所，给出了完美答案。位于瑞士苏黎世市西北部的Würenlingen幼儿园(138页)项目就提出了这个非同寻常的设计要求，目的是要解决周边城市环境空间严重缺乏的问题。首先，该设计运用了一种度量结构，以优化停车场；然后运用带有预应力梁的混凝土结构作为整个项目框架，架构出一座"混合建筑"。冷灰色的混凝土突出了整个结构的强度；而温暖的松木则标志着室内分隔构件和覆层的轻盈。

这个只有一层的幼儿园规模小巧、空间舒适，与可爱的小用户们建立起了融洽友好的关系。正如设计师所说，它就像一个"帐篷屋"，阳光能够照射进内部空间，而由混凝土模板建成的像窗帘一样的外立面赋予该建筑清晰的纹理和特色。室内空间之间物理联系和视觉联系的设计，不但保证了视线的通透，还使空间有了不只是单个房间的感觉。达到如此效果要归功于如下设计：室内走廊两侧是大型的透明玻璃墙；教室之间的混凝土梁上有大型圆形洞口；滑动门为整个空间开启了多种可能。

Completed in 2018 by London-based architects MUMA, the Storey's Field Center and Eddington Nursery (p.150) is located in Eddington, on the North West Cambridge Development, an area urbanized in the recent years to respond to Cambridge's urban pressure. It combines two types of architectural programs, a public infrastructure for the community center with a nursery for hundred children, clearly dividing functions and entrances. A square-plan organizes the volumes around a central "landscaped courtyard" that suggests a large cloister. Three wings of the building are dedicated to the nursery program, with a circulation opened over the central courtyard. Here, the playground is a "play garden", as architects named it, where benches perform playful shapes and the vegetation controls the views. Daycare-staff facilities are clearly separated from the children's areas. The circulation, peripheral to the patio and protected with a canopy, connects the nursery area and acts as a transition space – activities can happen here. Children's classrooms have generous surfaces and are designed as a house-like shape. It is a familiar space full of light and free enough to appropriate easily. Wide window-walls allow the garden to come inside. The sleeping-room brings the outside world discreetly, by a small round windows constellation, where kids can nurture their curiosity. Their senses are stimulated: some windows were drawn as familiar shapes with strong colors; bricks, stones, wood and other materials offer textures to experiment and touch; sound is absorbed by a patterned ceiling in the classrooms; the smell of the vegetation in the garden seems to penetrate the interior space.

The remaining wing, hosting the community center, turns to the outside of the building. Over the interior, the architects projected enclosed "private" patios that allow the green area to enter the space without invading the children's playground. This project expresses how architecture can be rational and geometrically sober and at the same time playful and stimulating.

How to combine two antagonist programs in one building, such as a kindergarten and an underground car parking? This was the task proposed through a competition, that the two Zürich-based offices Malte Kloes Architekten and Estrada Reichen Architekten answered. The Kindergarten Würenlingen (p.138) in the Northwest of Zürich, in Switzerland, has this unusual briefing with the purpose to solve the practical lack of space issues of the surrounding urban context. Firstly it imposes a metric structure in order to optimize the parking lots. A concrete structure with pre-stressed beams was projected to shelter the entire program and configure this "hybrid building". The cold grey concrete highlights the strength of the structure; the warm pinewood marks the lightness of the interior partitions and cladding.

With a small scale and cozy spaces, this one-floor kindergarten creates a friendly relation with its small users. As the authors refer, it resembles a "tent-house" where daylight is invited to come in and a curtain-like facade – molded by the concrete formwork – gives texture and character to the building. The physical and visual connections between the interior spaces allow visual permeability and the space to exceed the singular room: transparent walls over circulation; large circular

日本的MUKU托儿所(178页),由总部位于东京的Tezuka Architects建筑师事务所设计,是一个功能单一的建筑群项目。因为该项目的初衷是解决一家工业生产公司职员子女的入托问题,所以此托儿所就建于这家工厂附近。项目的挑战在于如何在工业区为孩子们营造一个激发孩子们灵感的优美环境,同时还能直接服务于当地社区。整个建筑群处在统一的封闭区域中,规划得如同公园,由一个个圆形单层模块结构组成,如同花园中一个个凉亭。每个亭子内都是一个房间,都拥有其特定功能。这些由木头和玻璃组成的圆形房屋整体看上去就像是一座小城市,孩子们在房屋里面玩耍、活动、探索、发现和学习。亭子内部是灵活而流动的自由空间,里面的家具会给孩子们提供一些线索,鼓励孩子们创造性地使用这些家具。亭子的通透性很好,四周是玻璃墙,这样的空间布局使其拥有360°的视野,孩子们也会360°无死角地完全处在看护人的视野之内。孩子们不但安全,而且时刻受到保护,可以自由地探索这个有趣公园的整个区域,富士山就在附近的景观中。托儿所还设计有小水池、人造地形以及多样化的植被等,大大丰富了孩童们的运动和感官体验。

还有一个项目是O6A地块幼儿园(164页),位于法国巴黎北部,是一栋集托儿所和幼儿学习设施以及住宅于一体的混合用途建筑。设计者是来自巴黎本土的SAM architecture和来自奥地利维也纳的Querkraft architekten两家建筑设计事务所。当时的工作背景是Clichy-Batignolles正在进行城市改造,需要为当地人口提供社会住房和配套设施。如今,大城市的建筑密度都有强制性要求,因此,住房项目被安排在两栋塔楼中,而两栋楼之间的空地设计了幼儿园和学校设施。整体布局保证阳光能最大限度地照射进学校内,同时充分利用城市地标和附近公园的景观。在居民区建造这样的学校设施可以减少接送孩子花在路上的时间,使父母有更多的时间和孩子在一起。学校设施的入口和住宅楼的入口是分开设置的,所以孩子们很安全。学校里有两个按年龄划分使用的室外操场,较低一层是专门供蹒跚学步的幼儿使用的,上面一层则是供少儿使用的,这样的设计给了孩子们更多享受室外活动的"自由"。

建筑师的设计理念表达了一种意愿,就是要规划一座灵活且容易适应不同情况的建筑。他们使用轻质材料做内墙,设计多功能的室内房间,还使用一些"次要"区域作为学习区,如走廊过道或门厅,从而创造出额外的活动空间。在设计过程中,建筑师们始终考虑到了未来用户的需求和空间的新功能。

openings in the concrete beams between classrooms; sliding-doors opening multiple possibilities for the space. MUKU Nursery (p.178) in Japan, by Tokyo-based studio Tezuka Architects, is a complex exclusively projected to serve this program. Promoted by an industrial production company to host their employees' children, it is located just by the factory. Here, the challenge was to design an inspiring environment for children within an industrial area, serving directly the local community. The entire complex is a unified closed area, planned as a park, with round-shaped single-story modules – like garden pavilions – that accommodate the program. Each pavilion contains a room with a specific function. The ensemble almost stages a small city, of wood and glass round houses, where children are encouraged to take action, discover and learn. Their interior is a flexible and fluid free space, where furniture gives some clues to users, pointing to creative appropriations. The transparency of the pavilions and their spatial layout with glass-walls around, allow a 360° view, where children are always being spotted by supervisors. Secure and protected, they are free to explore the entire area of this playful park, with the Mount Fuji in the nearby landscape. Features like a small water pond, a modeled terrain area or the diversified vegetation, enrich children's motion and sensorial experiences. Another project presented in this part, O6A Lot Kindergarten (p.164), brings the combination of mixed uses such as a housing complex with a daycare and learning infrastructure for young children in the North of Paris, France. The authors, SAM architecture from Paris and Querkraft architekten from Vienna, Austria, worked in the context of the Clichy-Batignolles' urban regeneration, where social housing and equipment were identified as needs for local population. Today densification is mandatory for large cities, therefore the housing program is hosted in two towers, which frame the kindergarten and school infrastructure, at the ground level. The general layout maximizes the sunlight entrance to the school and also takes advantage from its sights over urban landmarks and the public park. Providing this kind of equipment in a residential neighborhood enables to reduce travelling time, and give parents some more time to spend with their kids. Children are safely welcomed in the facility, as the entrances are separated from the housing blocks. Two exterior playgrounds, divided by ages, give children the possibility to enjoy exterior "freedom". The lower level is dedicated to toddlers and the upper floor to young children.
In their concept, architects expressed the will to plan a building that would be flexible and easy to adapt. They used light material for interior walls and proposed multi-functional rooms. Also, the idea of using "secondary" areas for learning, such as circulations or foyers, would create extra room for activities. Future users' requirements and new functions for the space were thoughts that have always been present during the design process.

1. Herman Hertzberger, *Space and Learning: Lessons in Architecture 3*, (Rotterdam: 010 Publishers, 2008, p.111).
2. Ibid. p.9.

Würenlingen幼儿园
Kindergarten Würenlingen

Malte Kloes Architekten + Estrada Reichen Architekten

瑞士北部城镇Würenlingen为孩子们建造的带公共地下停车场的类似帐篷结构的混凝土空间
Tent-like concrete space for kids with underground public car park in the northern Swiss town of Würenlingen

该设计竞赛的挑战在于如何设计一座能将地面层幼儿园和一个地下公共停车场紧凑结合起来的建筑。这一任务非同寻常，最终设计成的是一座综合建筑，该建筑满足了两种截然不同的功能要求，是一个超常规的建筑结构。幼儿园楼层平面设计以最小宽度18m为基础，这个宽度是地下停车场进行有效布局所要求的尺寸。为了使停车场结构天花板的负荷最小化，设计师将地面层幼儿园的屋顶设计成自由式跨越结构：预应力混凝土梁与大块混凝土板连接在一起，顺着室外坡屋顶的形状设置，从而将大部分的垂直荷载分散到外立面上。

同时，简洁的屋顶结构使人在屋子里感觉空间开阔，像大帐篷一样的室内空间通过屋顶中心细长的天窗采光，光线充足。混凝土梁与木质的非承重内置结构结合，将幼儿园建筑内部分成三个隔间，这三个隔间通过安装在沿街立面上的大型滑动门连接起来。混凝土梁上有大大的圆形洞口，在视觉上将室内主要空间和隐藏在混凝土梁之间舒适的洒满阳光的夹层连接起来。

出现在设计竞赛阶段的"帐篷式房屋"的比喻也影响了建筑外观的设计。在施工过程中，混凝土模板被覆盖了一层波浪形弹性材料，从而使裸露的混凝土立面拥有了精致的像窗帘一般的浮雕图案，有意与建筑的材料整体性形成反差。

幼儿园的结构概念保证了空间的开阔、灵活，没有设置起结构作用的室内墙体。此概念之所以能够实现，是因为建筑师采用了这样的设计：混凝土板被放置在四根梯形和两根矩形的预应力混凝土梁上，混凝土梁由双层混凝土外墙的内层承重墙支撑。同时，预应力梁也成为建筑空间概念的一部分，起到分割空间的作用，将幼儿园分割成相对独立的三个单元。

地下车库与地上幼儿园的所有结构构件都是采用现浇混凝土精心建造而成的。幼儿园房间里的一体化承重混凝土结构清晰可见，混凝土模板的图案以及模板震动留下的痕迹都经过精心设计和协调。在室内，可见的混凝土表面与非承重的木墙和固定装置互为补充。

幼儿园房间的特色是安装有玻璃墙，玻璃墙朝向可用作公共花园的室外空间开放。这一设计既创造了一种室内与室外环境之间的联系，又为室内提供了自然光照。

The challenge of the project competition was to design a building that could compactly combine a ground floor kindergarten with a public underground car park. This unusual task led to the design of a hybrid building that meets two very different programmatic requirements with an unconventional building structure. The floor plan of the kindergarten is based on the minimum width of 18m required for an efficient layout of an underground car park. In order to minimize the loads on the ceiling slab of the parking structure, the roof of the kindergarten on the ground floor was designed as a free spanning structure: prestressed concrete beams in conjunction with an extensive concrete slab follow the shape of the exterior hip roof, distributing the majority of all vertical loads to the exterior facades.
At the same time, the concise roof structure creates atmospheric, tent-like spaces on the inside, which are flooded

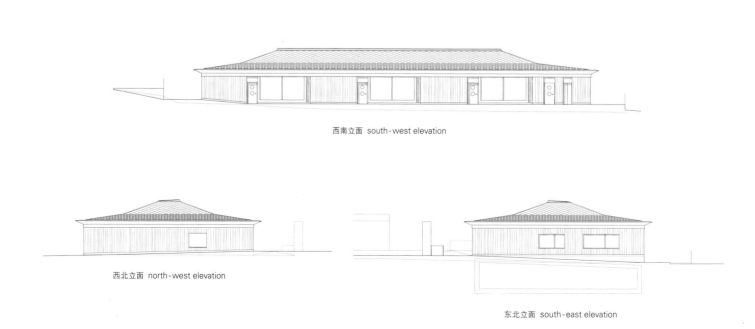

西南立面 south-west elevation

西北立面 north-west elevation

东北立面 south-east elevation

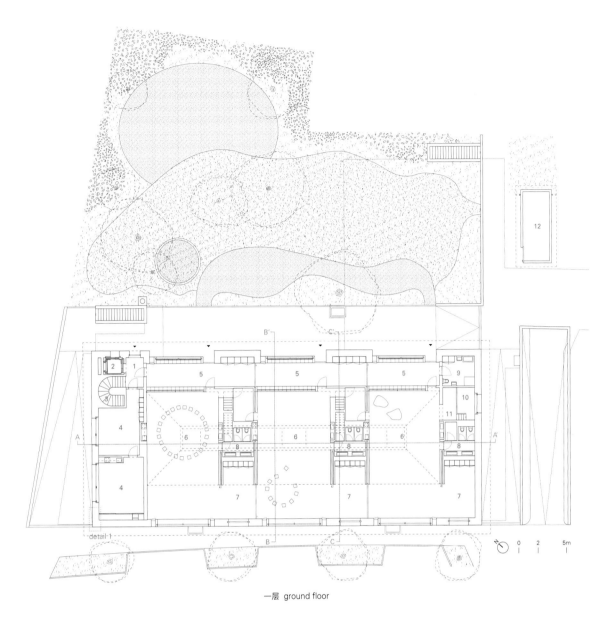

一层 ground floor

1. 入口 2. 阁楼 3. 电梯井 4. 员工室 5. 入口大厅/衣帽间 6. 教室
7. 小组活动室 8. 卫生间 9. 残疾人卫生间 10. 游戏室 11. 走廊 12. 花园小棚屋 13. 停车场

1. entrance 2. loft 3. stairwell 4. staff room 5. entrance hall/cloakroom 6. classroom
7. group room 8. toilet 9. disabled toilet 10. playroom 11. gallery 12. garden shed 13. parking

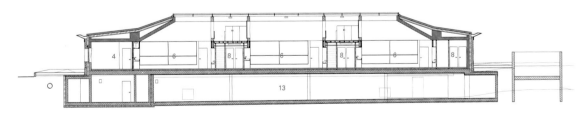

A-A' 剖面图 section A-A'

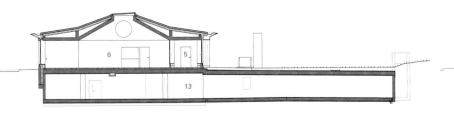

B-B' 剖面图 section B-B'

C-C' 剖面图 section C-C'

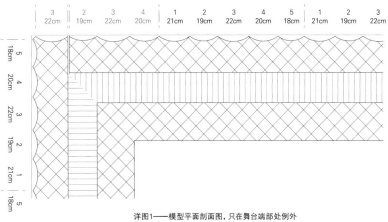

详图1——模型平面剖面图，只在舞台端部处例外
detail 1 _ section of mold plan with one exception at the end of a stage

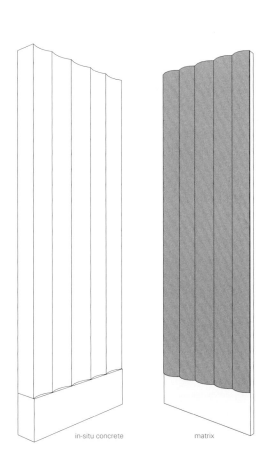

模板标准模型，悬垂部分与基座连接
standard mold on formwork panel
with overhang for connection to pedestal

in-situ concrete　　　matrix

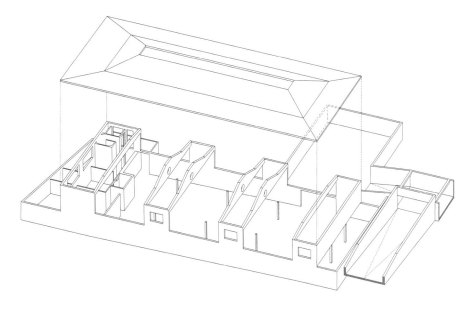

项目名称：Kindergarten Würenlingen / 地点：Würenlingen, Switzerland / 事务所：Malte Kloes Architekten + Estrada Reichen Architekten /
项目团队：Malte Kloes, Christoph Reichen, Andreas Lochmatter, Sven Rickhoff, Daniel Schürer / 景观建筑事务所：USUS Landscape Architecture /
结构工程公司：Caprez Ingenieure AG / 暖通空调工程公司：Mettauer AG / 建筑物理：Steigmeier Akustik + Bauphysik GmbH /
电气工程：HKG Engineering AG / 照明工程：Reflexion / 客户：Municipality of Würenlingen, Canton Aargau, Switzerland /
用途：kindergarten / 用地面积：2,416m² / 建筑面积：732m² / 总楼面面积：1,951m² / 造价：7.1 Mio. CHF / 设计时间：2015.2—2018.8 /
施工时间：2017.2—2018.9 / 摄影师：©Lukas Murer (courtesy of the architect)

with daylight via generous linear skylights in the center. The concrete beams, in combination with the wooden, non-load bearing built-in structures, divide the interior into three kindergarten compartments, which are connected via large sliding doors on the street facade. Large, circular openings in the concrete beams allow for visual connections between the main spaces and the cozy, sky-lit mezzanine hide-outs between the beams.

The analogy of the "tent house", which emerged during the competition phase, also shaped the design of the external appearance of the building; by cladding the formwork panels with wavy elastomer matrices during construction, the outer surface of the exposed concrete facade inherited a delicate, curtain-like relief, deliberately contrasting with the monolithic materialization of the building.

The structural concept of the kindergarten stipulates open, flexible spaces without structurally activated interior walls. This is rendered possible by fitting a concrete slab into four trapezoidal, and two rectangular, prestressed concrete beams which rest on the inner, loadbearing wall of a double-leaf external concrete wall. At the same time, the prestressed beams are integrated into the spatial concept by functioning as partitions separating the three kindergarten units from one another.

All structural elements of the underground garage and the overlying kindergarten were crafted with cast-in-place concrete. The loadbearing, monolithic concrete structure in the kindergarten rooms is left visible, and the formwork pattern, as well as the associated shocks of the formwork panels, are carefully planned and coordinated. Inside, the visible concrete surfaces are complemented with non-loadbearing timber walls and fixtures.

Kindergarten rooms feature glazed walls that open onto outer space which will be used as a public garden. This creates a link to the outside surrounding, while providing natural light.

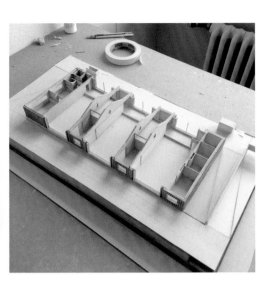

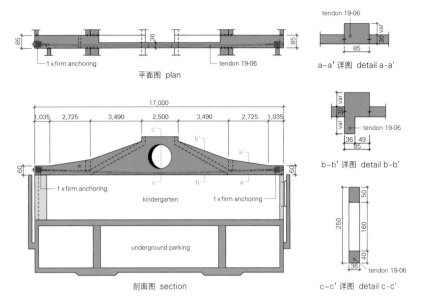

梁剖面，梁将地下车库与幼儿园结合在一起，地下车库的结构或网格与幼儿园并不对应
section of the beam that brings together structures or grids of the underground garage and the kindergarten that do not correspond

Storey's Field中心与爱丁顿托儿所
Storey's Field Center and Eddington Nursery

MUMA

Storey's Field中心与爱丁顿托儿所
A nursery and community center in Cambridge characterized by patterned brickwork and sheltered courtyard

Storey's Field中心与爱丁顿托儿所服务于爱丁顿新社区，是剑桥西北发展规划区的一部分。托儿所位于剑桥大学附属小学旁边，能容纳100个孩子，其三面都是景观庭院，第四面是社区中心，可以满足社区成年人的需求。最大的体量是可以容纳180人的大厅，它成为整个建筑项目的视觉中心和功能中心。

大厅的主体量可以满足各种声学要求，音响效果可以根据不同的活动进行调节，从室内音乐演奏到电影放映，活动多种多样。大厅的高度对于厅内被动通风、声学衰减空间效果至关重要。大厅被设计成方形，浅色的木梁与灰褐色的砌砖和谐相融。室内建有螺旋楼梯，成为通向三层高空间上层的通道。

社区中心拥有多功能聚会场地，涵盖了像举行婚礼、追悼会、尊巴舞课或者太极等这样的活动。另外，中心还设有会议室、办公室、厨房和储藏间。

门厅自带花园，还有一个带围墙的花园装点着整个建筑综合体的东南角。此处的墙壁上设计了圆形孔洞，创造出变幻莫测的光影效果，十分有趣。

建筑师通过与社区咨询和协商，确定了设计任务书，整个建筑项目需要建造三个规模不一的主要空间（体育设施、表演艺术空间、幼儿园）。鉴于社区居民可以利用附近的体育设施，所以整个项目把重点放在了表演艺术空间的设计上。

庭院为托儿所的孩子们提供了可以躲避风雨的嬉戏空间，虽没有篱笆围挡，但同样为孩子们提供了安全保障。庭院里的回廊成为教室的外部走廊，同时也是孩子们的嬉戏之所。教室没有走廊，直接与花园相接，在教室中不但能看到旁边学校的操场，还能感受到穿堂风在教室里习习吹过。回廊设计与大学学院建筑特色遥相呼应，向其致敬。

人们可从建筑四周每个角度来审视这座建筑，每一个立面都经过精心的设计和构思，有图案精美的砌砖造型，有嵌入式、上有遮挡的入口，还有精雕细刻的石头座椅。在整个设计中，除了一些妙趣横生的设计元素，设计的焦点是与花园的联系，与窗户远处风景的联系。大大的室内窗户非常醒目，有三角形的，有圆形和方形的，窗台很宽很深，可以当座位用。孩子们透过这些大窗户和小小的舷窗可以窥探外面的世界。天花板上密密麻麻的小圆孔也是这些窥探游戏的一个缩影。

面积较大的剑桥西北发展规划区项目包括一所小学、研究生宿舍、大学和学院职工的住处以及一座市场。这个项目由剑桥大学委托设计建造。

The Storey's Field Center and Eddington Nursery serve the new community of Eddington, on the North West Cambridge Development. Located adjacent to the University of Cambridge Primary School, the nursery building, which accommodates 100 children, is arranged around three sides of a landscaped courtyard. On the fourth side, the community center addresses the needs of the adult community. The largest volume, the 180 person main hall, forms the visual and programmatic center of the complex.

The main hall's volume allows for variable acoustics that can be adjusted to suit events ranging from chamber music to film screenings. Its height is critical to achieving a passively ventilated, acoustically attenuated space. The hall is designed with square, pale-colored timber beams to complement the sandy pallor of the brickwork. A spiral staircase provides access to the upper floors of the triple-height space.

与剑桥球场规模相当的学院
relative scale of Cambridge courts

彭布罗克学院 Pembroke College | 三一学院 Trinity Hall | 莫德林学院 Magdalene College | 冈维尔与凯斯学院 Gonville & Caius College | 基督圣体学院 Corpus Christi College | 皇后学院 Queens College | 爱丁顿社区中心与托儿所 Community Center & Nursery, Eddington

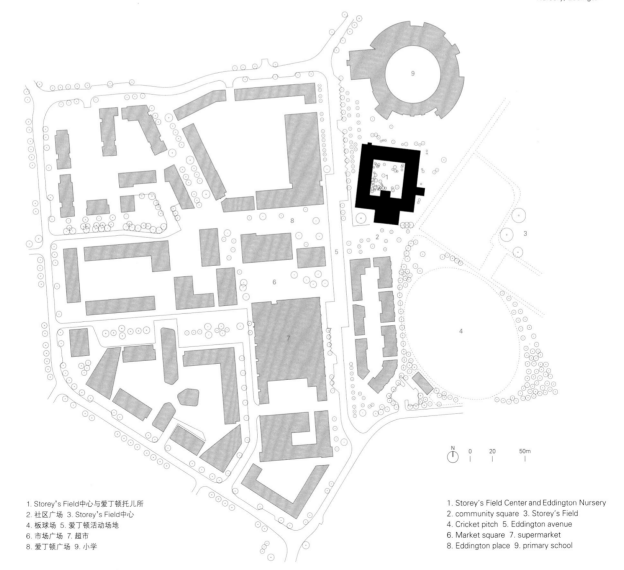

1. Storey's Field中心与爱丁顿托儿所
2. 社区广场 3. Storey's Field中心
4. 板球场 5. 爱丁顿活动场地
6. 市场广场 7. 超市
8. 爱丁顿广场 9. 小学

1. Storey's Field Center and Eddington Nursery
2. community square 3. Storey's Field
4. Cricket pitch 5. Eddington avenue
6. Market square 7. supermarket
8. Eddington place 9. primary school

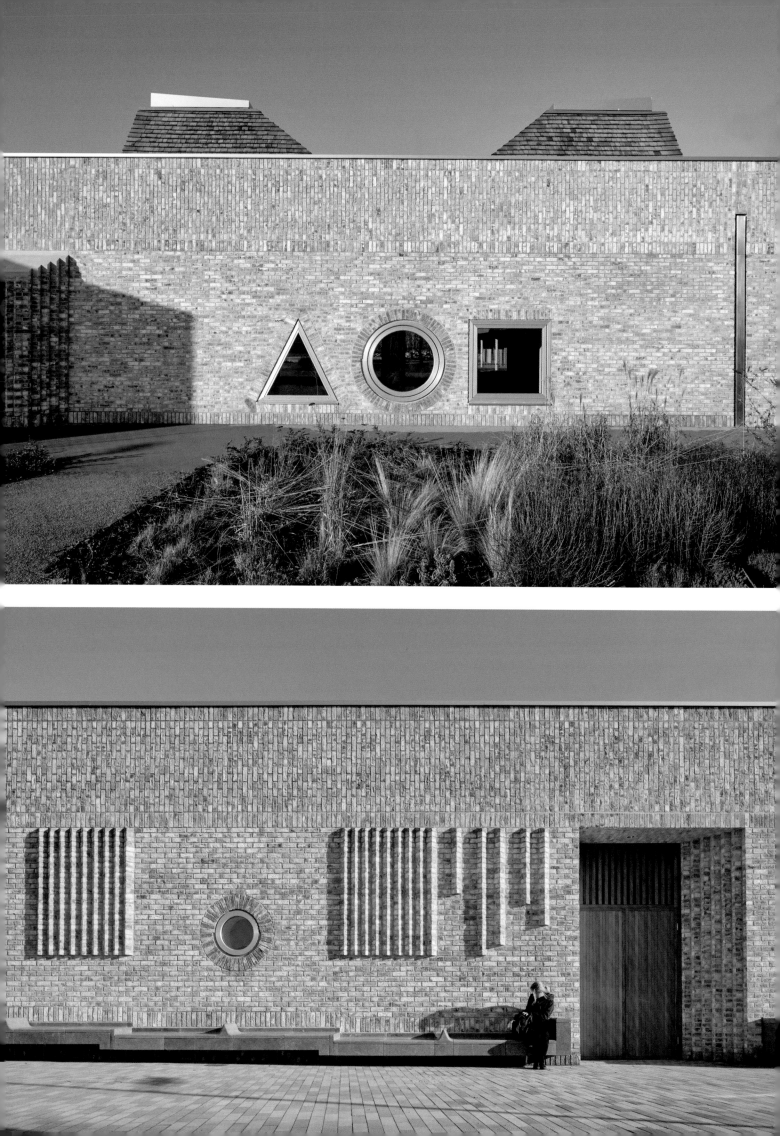

托儿所
1. 门厅与接待处
2. 婴儿车存放处
3. 经理办公室
4. 家长会面室
5. 游戏花园
6. 回廊
7. 教室入口大厅/衣帽间
8. 教室
9. 换尿布房间
10. 儿童卫生间
11. 公共房间
12. 睡眠房间
13. 热奶厨房
14. 档案室
15. 总办公室
16. 员工室
17. 厨房/储藏间
18. 杂物室
19. 员工衣帽间/储物柜
20. 信息技术室
21. 储藏室

社区中心
22. 门厅&接待处
23. 办公室
24. 厨房/储藏间
25. 门厅花园
26. 小多功能室
27. 大多功能室，带储藏间和小厨房
28. 带围墙花园
29. 主厅，带座椅和设备储藏间
30. 主厅花园
31. 卫生间
32. 储藏室

共享设施
33. 垃圾存放处
34. 电气设备室
35. 淋浴室与储物柜

nursery
1. foyer & reception
2. pram store
3. managers office
4. parents' meeting room
5. play garden
6. cloister
7. classroom entrance lobby/coats
8. classroom
9. nappy changing room
10. children's WCs
11. communal room
12. sleep room
13. milk kitchen
14. archive
15. general office
16. staff room
17. kitchen & store
18. utility room
19. staff coats & lockers
20. IT
21. store

community center
22. foyer & reception
23. office
24. kitchen & store
25. foyer garden
26. small multi-purpose room
27. large multi-purpose room – with store & kitchenette
28. walled garden
29. main hall - with chair & equipment stores
30. main hall garden
31. WCs
32. store

shared facilities
33. refuse
34. plant & electrical
35. shower room & lockers

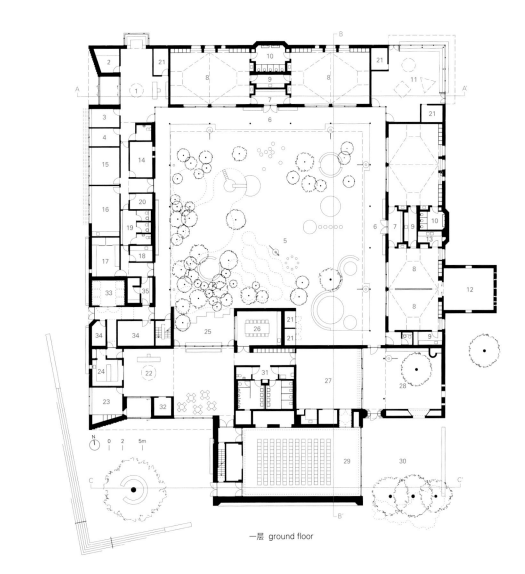

一层 ground floor

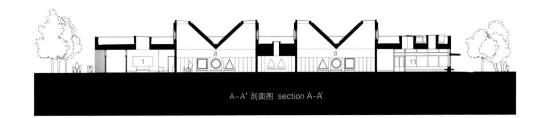

A-A' 剖面图 section A-A'

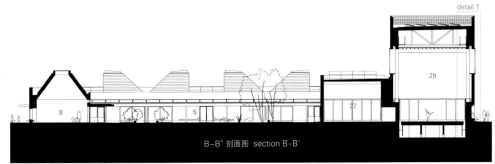

B-B' 剖面图 section B-B'

项目名称：Storey's Field Centre and Eddington Nursery / 地点：Cambridge, England, UK / 事务所：MUMA / Structural engineer, MEP consultant, building physics, acoustician (nursery), fire engineering, civil engineering: Aecom / 项目经理：Turner & Townsend / 工料测量师、成本顾问：Gardiner & Theobald / 景观顾问：Sarah Price Landscapes / 剧院与声学顾问 (社区中心)：Sound Space Vision / 结构工程公司：FMDC Ltd / 照明设计：Lumineer / 入口设计顾问：Center For Accessible Environments BREEAM顾问：NHBC / CDM协调：principal designer – Faithful + Gould (previously CDM Coordinator) / 建筑检察院员：3 shared services – Cambridge City Council 总承包商：Farrans Construction Ltd / NEC监理：Calfordseaden / 客户：University of Cambridge / 用地面积：3,425m² / 总楼面积：2,248m² / 造价：£8.28M 竣工时间：2018.1 / 摄影师：©Alan Williams (courtesy of the architect)

The community center provides a multi-purpose gathering place, accommodating marriages, memorial services, Zumba classes or Tai Chi. It also includes meeting rooms, offices, kitchens and storage space.

The foyer has its own garden, and another walled garden makes up the southeastern corner of the complex. Here, circular holes have been incorporated into the walling to create interesting interplays of light.

The brief, developed through community consultation, called for three main spaces of varying scale. As the community already benefit from nearby sports facilities, the brief developed a strong focus on the performing arts instead.

The courtyard offers sheltered play for the nursery children, providing security without fences. A cloister allows external circulation to the classrooms and covered play. With no corridors, the classrooms can engage directly with the garden and also have views towards the adjacent school playground, while benefiting from cross ventilation. The cloister is a contextual nod to the characteristic architecture of the university colleges. The building can be viewed in the round, and each facade is carefully composed and articulated, with patterned brickwork, inset sheltered entrances and carved stone seating. Throughout there is a focus on contact with gardens and views to the landscape beyond, along with playful elements.

Large interior windows, in triangles, circles and squares, have deep sills which function as seats. These, and smaller porthole windows, allow children to peek outside. Round perforations in the ceilings mirror these games with shapes.

The wider North West Cambridge Development includes a primary school, post graduate student accommodation, university and college worker homes, and a market. The project was commissioned by the University of Cambridge.

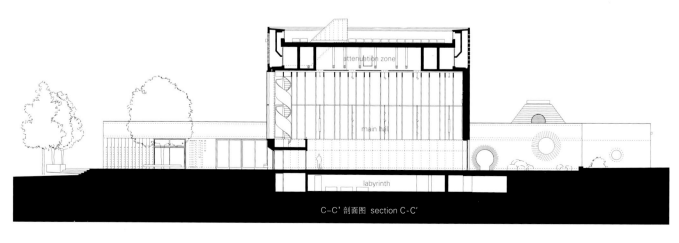

C-C' 剖面图 section C-C'

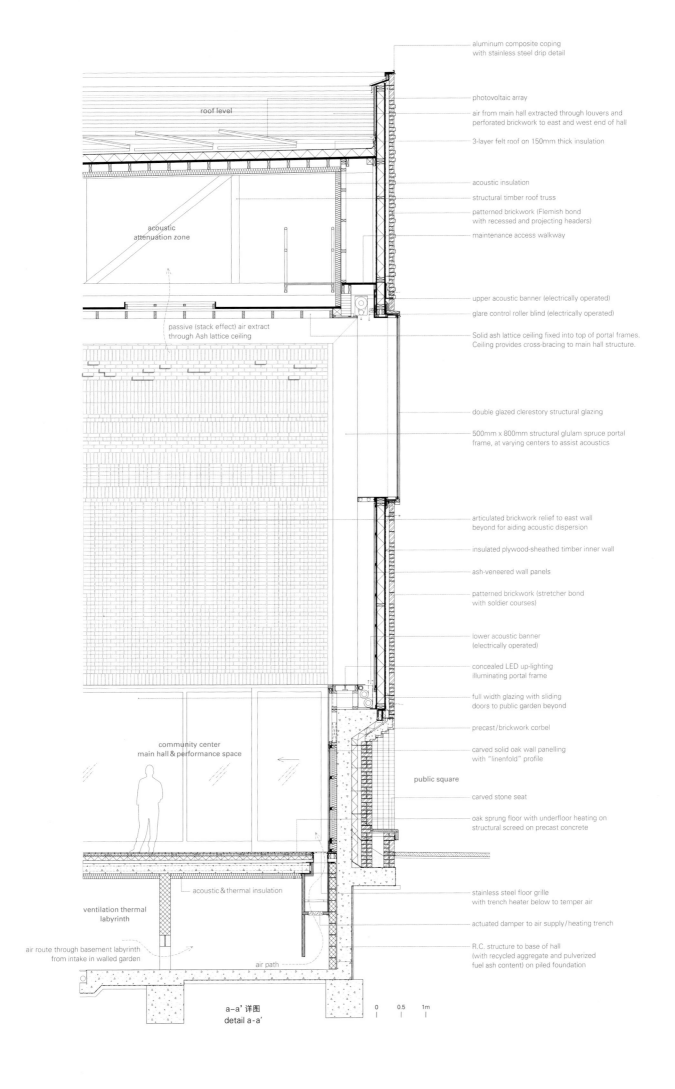

O6A地块幼儿园
O6A Lot Kindergarten

sam Architecture + Querkraft

隶属巴黎某公益住房的幼儿园展示了其木结构的重要性
A kindergarten belonging to a Parisian social housing reveals its materiality with wooden structure

Clichy-Batignolles是一个雄心勃勃的城市改造区，位于巴黎第17区，占地超过538 000m²。这个历史悠久的地段拥有通往圣拉扎尔火车站的铁路线和封闭的环城公路，该项目旨在为提高周围建成环境的城市密度尽一份力。

环绕马丁·路德·金公园建造的O6A地块项目包括127户公益住房、一家幼儿园和一家日间托儿所。该项目最重要的目标之一是幼儿园和日托中心的户外操场都能够拥有尽可能多的空间和阳光照射。建筑空间之间的孔隙和视野尤为重要。最终的建筑体量设计效果就是理性考虑了项目的功能和场地的限制条件。

设计最终的选择是增加居住区的密度，建造两栋高达50m的公寓楼。这两栋塔楼建在该地块的南北对角线上，这样不仅两个操场都拥有最大限度的阳光照射，而且中午时分，塔楼的影子只能遮挡到幼儿园的屋顶。该项目力图朝向公园方向开放，塔楼之间的空间创造了街道与公园之间的通透视野，公园里树影婆娑，树木葱茏，在幼儿园即可将其尽收眼底。

幼儿园和托儿所的设计目标一致，即从被人们普遍忽视的、被认为是不重要的"次要"空间中设计出尽可能多的有价值的可用空间。比如，走廊变成额外的房间，教室的前厅用作小组活动空间等。设计者的目标就是最大化地开发空间的潜能，使每个房间都具备多种功能。每个房间都拥有潜能是对设计概念的支持，这一概念就是每个空间都可以随时用来举办各种活动。幼儿园也可以在孩子们放学后成为社区举办活动的地方，一些空间可以用来供大家使用。其中一个舞台的设计，就是将人来人往的大厅转换成灵活的活动空间。在这儿，可以远观蒙马特山（译者注：Montmartre，位于巴黎的一座小山丘）独特的景色。该建筑的设计意图从长远来看就是要适应性强，灵活多变。采用柱梁结构和轻型墙体就是实现这一设计策略的一部分。这样的建筑可以在未来不改变整体建筑风格的前提下，进行重新布局，发挥新的室内功能。建筑师们竭力尊重未来用户的需求进行设计，但在设计愿景上绝不妥协。

材料的使用问题对于建筑来说至关重要，但建筑外立面及其装饰往往限制了其使用。该项目旨在透视许多建筑在材料使用方面的对立性，主要体现在结构材料或者技术材料与那些用于伪装或遮挡它们的材料之间的冲突上。这些装饰性的材料往往只有薄薄一层，使用寿命较短。

而该项目打破了依赖时尚的装饰模式，取而代之的是在建筑结构的展示中强调材料的原始质地。材料的美感通过这种真实可靠的方法得到了增强。技术构件没有被隐藏起来而是显露在外，展示了其功能性。

该建筑的主要设计理念是要在构造中追求真实。幼儿园和托儿所从里到外都是木结构。在建筑外部，木覆层和法国落叶松木框架演绎出了木结构的灵魂。这些材料也代表了建筑对结构的需求，就如同建筑的外观一样。因此，影响建筑风格的元素是建筑的外观。要想改变建筑构造，就要改变结构部件，否则是不可能的。

The ambitious urban regeneration zone of Clichy-Batignolles, in Paris' 17th arrondissement, covers over 538,000m² of land. The historic site is served by train lines leading to Saint Lazare train station and the close beltway; the intention for the project is to contribute to a greater urban density in the built environment of the neighborhood.

Lot O6A, constructed a round Martin Luther King park, includes 127 social housing facilities, a preschool and a day nursery. One of the overriding goals is to obtain as much space and sunlight as possible for the outdoor playgrounds of the school and daycare center. The question of porosities and views created between the spaces is vital; the final volumetric design is the direct result of a rational consideration of the programmatical and site constraints.

The choice was made to densify the housing areas, grouping apartments into two buildings of up to 50m high. These towers, on the north and south diagonal of the site, allow maximum sunlight for both playgrounds, and at noon the shadow from the tower covers only the roof of the kindergarten. The project is intentionally as open as possible towards the park, with the space in between the towers creating views from the street to the park, with its foliage and trees also visible from the school.

Both the school and the kindergarten were conceived with a common goal: to create as much valuable, usable space as

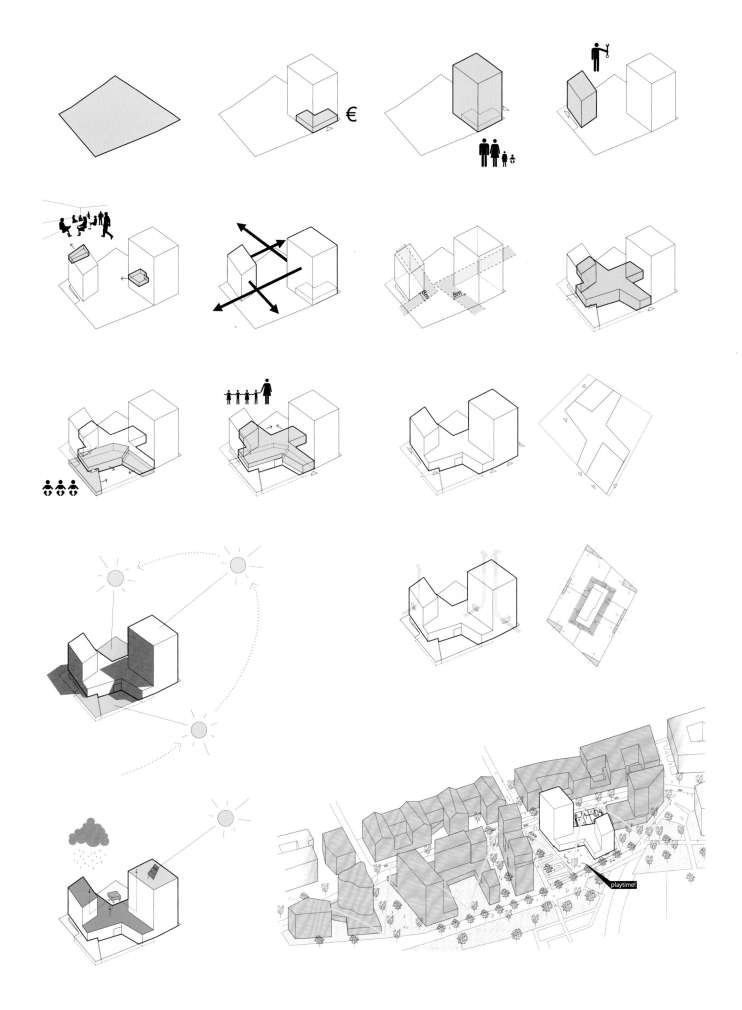

1. main structure's column, larch
2. beam main structure, larch
3. vapor barrier 18mm
4. window joinery, larch
5. thermal insulation 12mm
6. OSB panel 12mm
7. rain screen 0.02m
8. exterior awning
9. upper aluminum cover strip
10. lower aluminum cover strip
11. pine tree battens (22x40)
12. wooden cladding: larch
13. OSB board
14. parapet's aluminum cover strip
15. post's aluminum cover strip
16. anti splitting screw
17. outdoor larch slat
18. handrail
19. cover board, birch

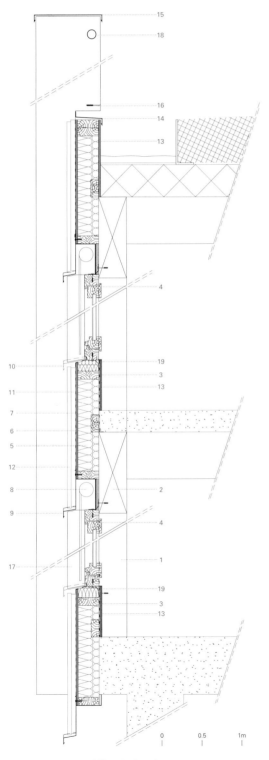

a-a' 详图 detail a-a'

公共设施
1. 学校入口
2. 物理治疗室
3. 运动室
4. 第一阅览室
5. 步行桥
6. 日托中心中央大厅
7. 学校食堂
8. 厨房
9. 日托中心入口

住宅
10. 储存空间
11. 公寓建筑入口
12. 自行车停放处
13. 住宅单元建筑入口

public facilities
1. school entrance
2. physiotherapy room
3. motricity room
4. first books
5. footbridge
6. daycare central hall
7. school canteen
8. kitchen
9. daycare entrance

housing
10. store
11. apartments building entrance
12. bike garage
13. housing units building entrance

一层 ground floor

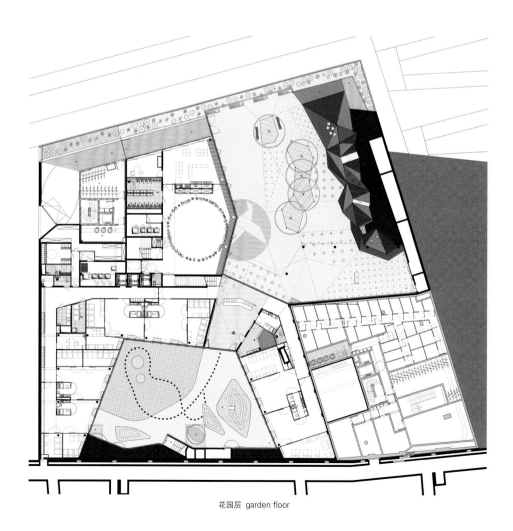

花园层 garden floor

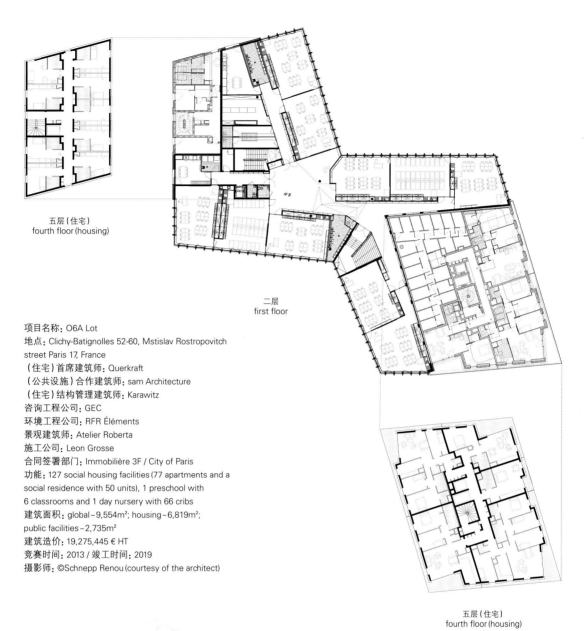

五层（住宅）
fourth floor (housing)

二层
first floor

五层（住宅）
fourth floor (housing)

项目名称：O6A Lot
地点：Clichy-Batignolles 52-60, Mstislav Rostropovitch street Paris 17, France
（住宅）首席建筑师：Querkraft
（公共设施）合作建筑师：sam Architecture
（住宅）结构管理建筑师：Karawitz
咨询工程公司：GEC
环境工程公司：RFR Éléments
景观建筑师：Atelier Roberta
施工公司：Leon Grosse
合同签署部门：Immobilière 3F / City of Paris
功能：127 social housing facilities (77 apartments and a social residence with 50 units), 1 preschool with 6 classrooms and 1 day nursery with 66 cribs
建筑面积：global – 9,554m²; housing – 6,819m²; public facilities – 2,735m²
建筑造价：19,275,445 € HT
竞赛时间：2013 / 竣工时间：2019
摄影师：©Schnepp Renou (courtesy of the architect)

1. 日托中心中央大厅 2. 步行桥 1. daycare central hall 2. footbridge
A-A' 剖面图 section A-A'

possible from areas which might ordinarily be overlooked as secondary. A corridor can become an extra room, the antechamber of a classroom could be useful for small group activities, and so on. The designers' aim is to maximize the potential of space, such that every room can enjoy multiple functions. The potential contained in each space supports the idea that each one could host various activities and moments.
It is intended that the school could also become a place for other community activities once the classes are done, opening out the spaces into welcoming places for others. The presence of a stage, framing a unique view of Montmartre, converts the room from a circulation hall to a flexible space for hosting events. The architecture is intended to be adaptive and flexible in the long term. The choice to showcase a structure made of columns and beams with light walls is part of this strategy; the building can be redefined in the future to host new indoor functions without changing its global architecture. The architects' have endeavored to design a place that respects the needs of its future users without compromising on design vision.
The issue of materials is essential to architecture, but all too

often limited to facades and ornamentation. This project aims to put into perspective the antagonism of many buildings, whereby a conflict exists between structural or technical materials and those materials used to disguise them, in a shallow, ornamenting layer with a short life span.

Here, the project breaks free from fashion-dependent ornamentation: instead the raw quality of the materials is underlined, by revealing the structure of the building. The material beauty is enhanced by the authenticity of this approach. Technical components are left visible, showcasing their functionality.

The main concept is to aim for truth in construction. The wooden structure of the kindergarten and school can be perceived both from inside and outside the building. On the outside, the spirit of the wooden structure is translated into wooden cladding and frameworks of French larch.

These materials represent a structural necessity for the building as much as its external appearance. As such, an element of architectural control remains over the appearance of the building since it becomes virtually impossible to change the aspect of the construction without damaging its structural parts.

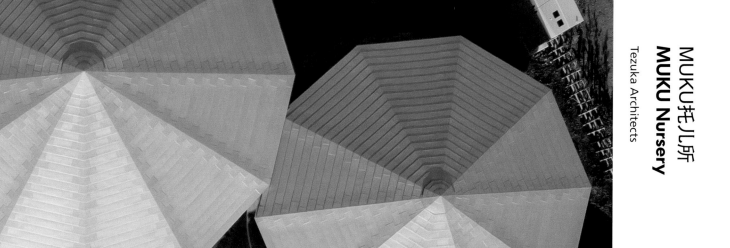

MUKU托儿所
MUKU Nursery

Tezuka Architects

日本幼儿园的气泡式平面设计使孩子们的本能运动尽在广阔视野范围内
Japanese kindergarten with bubble-like plan provides wide visibility of the children's instinctive movements

MUKU托儿所位于日本富士山地区,为一家生产便当盒的公司所有,托儿所就处在公司附近。该项目由公司主导建造,为了帮助那些孩子没人照顾的公司员工,不但是员工的福利,同时也服务于当地社区。

MUKU托儿所的平面与其说是圆形的,不如说是"泡泡"状。整个平面看起来就像数个泡泡缓缓地升至空中,彼此之间保持着最佳距离。每座泡泡建筑只有一个功能,内部没有墙体。像一个个由细胞器、线粒体和核糖体支撑起来的单细胞有机体一样的泡泡建筑是由家具和低矮的隔墙支撑的。

圆形平面设计总是受到诟病,因为它不易于布置家具,而且结构也太复杂。"我对此并无异议",建筑师说,但鉴于该项目的特殊需要,"我们发现圆形的规划比传统的模块化规划有更明显的优势"。

首先,每项功能的设置完全不受几何形状的限制。每个房间都可以像在图示中所分析的那样布局。房间的大小也可以自由调节。

其次,可视性极佳,这也是幼儿园和托儿所要考虑的最重要的要求之一。圆形的结构自然产生360°无死角的可视度。泡泡建筑之间的间隙可以让人从场地的一端看到另一端。办公室的视野很像是坐在船的驾驶台上的视野。

第三,圆形建筑也激励着儿童们无休止的、持续的本能运动。这些圆连接和创造了无限种可能的组合。

不过,这其中只有一个圆可以让孩子们径直穿过,它就是位于场地中央的小"水洼",孩子们在此嬉戏就好像在一个水碟里嬉戏玩耍一样。这里面没有台阶,坡度舒缓,只有30cm深。冬天,排干水碟里的水,小水洼就变成了孩子们可以在里面玩耍的空旷的地方。中央的"碟子"用混凝土浇筑而成,形成了平缓的坡度,很像海岸。

MUKU托儿所的主要结构为木结构。屋顶木椽结构的一个主要特征是:椽子呈放射状布置,并在中心点相交。中心点位置处的木块发挥了拱顶石的作用,以平衡放射状分布的结构,这样就不需要安装中央支柱了,从而创造出非常流畅的中央空间。椽子和中央木块之间的木作都是精心制作而成的,达到了不需要螺栓或销钉就能使用的程度。这样,整个结构的作用力就从中心点向边缘传递,然后再沿着周边支柱向下传递。

Located in Fuji, Japan, MUKU Nursery is owned by a company that produces bento boxes and is located within its vicinity. It was a company led initiative to subsidize the employees, helping with the lack of child care in order to contribute towards the employees' welfare, as well as to serve the local community. The plan for MUKU Nursery can best be described as "bubbly" rather than circular. The plan resembles bubbles, slowly rising in air, while keeping optimum distance between each other. Each architectural bubble has only one function. There is no wall inside; like a single cell organism supported by organelles, mitochondria and ribosomes, each bubble is supported by furniture and low partitions.

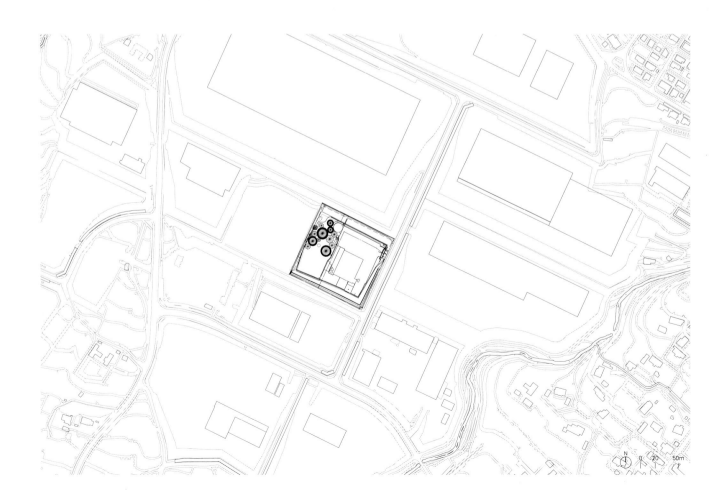

There are always criticisms about circular plans because it is not easy to arrange furniture and the structure is complicated. "I have no objection against those", says the architect, but for the needs of this particular project "we found there are significant advantages on this circle planning comparing to conventional modular planning".

First of these is the positioning of each function as completely free from geometrical restrictions. Each room can be located as it is analyzed on the diagram. The size of each room can be adjusted freely.

Secondly, the visibility is excellent, one of the most important requirements of a kindergarten and nursery school. The round shape brings 360 degree visibility naturally. The gap formed between the "bubbles" provides glimpses from one end of the site to the other. The view from the office resembles the view one gets from the bridge of a ship.

Thirdly, round shapes encourage the endless circular movement of children, continuous instinctive movement. These circuits interlink and create infinite combinations.

Yet there is only one circle which allows the children to cross it directly, a central "puddle", like a dish of water for the children to play in. Without steps, a gentle slope leads to a depth of only 30cm. In the winter, this water is drained and the pool becomes an empty plateau in which the children can play and process around. This central "dish" is cast in concrete to form a gentle gradient emulating a shore.

The main structures of MUKU Nursery are of timber construction. One main feature in the construction of the roof rafters is that they follow the direction of the radial form to meet at a central point. The wooden piece located at this central point acts as a keystone in order to balance the radial structure and to remove the requirement of a central column, thereby creating a fluid central space. The joinery between the rafters and the central wooden piece is finely crafted to the extent that no bolts or pins are used within this joinery. The forces are then transferred outwards from the central point and downwards along the peripheral columns.

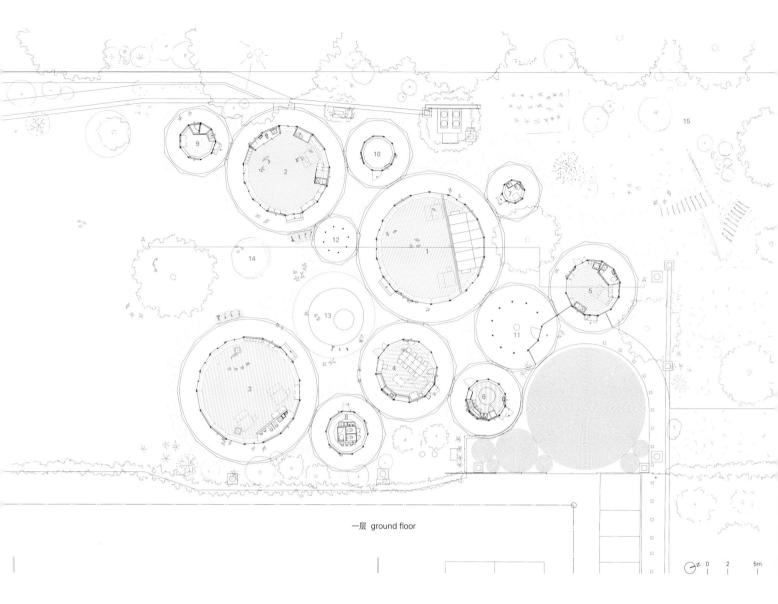

一层 ground floor

1. 多功能室 2. 1岁以下幼儿护理室 3. 2~3岁幼儿护理室 4. 办公室 5. 生病儿童护理室 6. 厨房
7. 普通卫生间 8. 卫生间 9. 更衣室与淋浴间 10. 储藏室 11. 入口 12. 走廊 13. 水上游乐场 14. 沙地游乐场 15. 假山游乐场
1. multi-purpose room 2. child care room for up to 1-year-old 3. child care room for 2 and 3-year-old 4. office 5. post-illness child care room 6. kitchen
7. general toilet 8. toilet 9. changing and shower room 10. storage 11. entrance 12. aisle 13. water playground 14. sand playground 15. mountain playground

A-A' 剖面图 section A-A'

项目名称：MUKU Nursery
地点：Shizuoka Prefecture Fuji City, Japan
事务所：Tezuka architects – Takaharu Tezuka Yui Tezuka, Kenta Yano
结构：Ohno Japan – Hirofumi Ohno, Satoshi Fujimoto
景观：10 kei – Keiichi Ishikawa
设施：Office May – Yousuke Hitomi, Minami Yuya, Oshima Masami
照明：Bonbori Lighting Architect & Associates,inc – Masahide Kakudate, Toshio Takeuchi
施工：Sato Construction – Shingo Karuishi, Kojima Takeru
艺术指导：Marukin Ad – Okada Hideaki, Mai Kanno, Yamagishi Erika
用途：nursery
用地面积：5,497.99m²
建筑面积：537.19m²
总楼面面积：403.51m²
结构：wooden
设计时间：2016.11—2017.6
施工时间：2017.7—2018.2
摄影师：©Kida Katsuhisa (courtesy of the architect)

1岁以下幼儿护理室平面图
1-year-old childcare room plan

2岁以下幼儿护理室平面图
2-year-old childcare room plan

办公室平面图
office plan

婴儿室/办公室
(0~1岁幼儿) 55m²
baby room/office
(0, 1-year-old child)
55m²

幼儿护理室
(2岁幼儿) 45m²
childcare room
(2-year-old child)
45m²

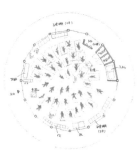

幼儿护理室
(2~3岁幼儿) 85m²
childcare room
(2, 3-year-old child)
85m²

生病儿童护理室23m²
sick child care room
23m²

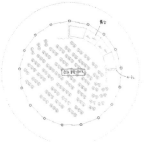

多功能室91m²
multi-purpose room
91m²

服务室13m²
serving room
13m²

办公室
药房44m²
office
medical office
44m²

卫生间10m²
toilet 10m²

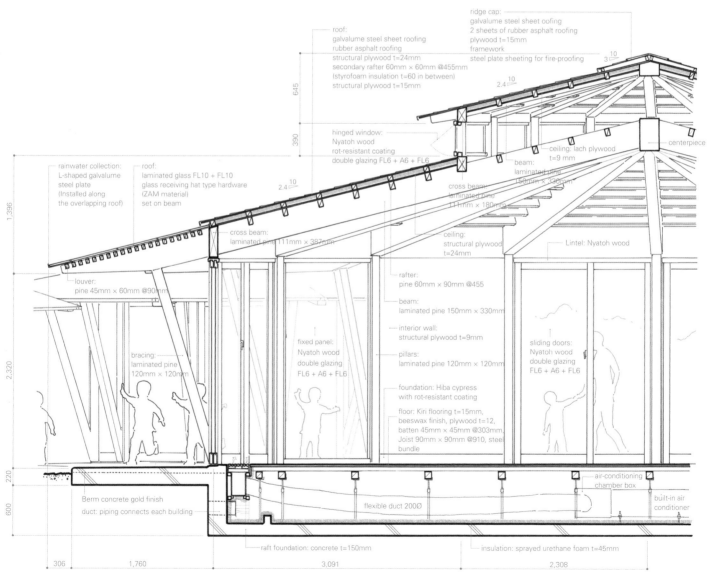

详图1 detail 1

"如水一样，我的朋友"

"Be Water,

水元素是建筑的经典伴侣：水拓展了建筑的意义，仿佛让人如置身一场思考的游戏中；水也能抑制建筑的形式主义，使建筑空间更令人愉悦。

与建筑相比，水更能够展现其流动性，展现其形式的千变万化，其力量既能势不可挡，也能使人平静放松。从这个意义上来说，它成功地改变了建筑的一大主要特征，即建筑的静止性。得益于水这种液体元素的存在，建筑的静止性被赋予了潜在的可变性。

本文展示的例子讲述了水和建筑之间的关系，建筑因水元素的融入而外形多变，弯曲

The element water is a typical "partner" of architecture: it amplifies its meanings, as in a game of reflections, but it is also capable of dampening its formalisms, of making the built space more joyful.
Compared to architecture, water is able to show itself in its fluidity, in its ability to take on any form, to be disruptive in its power, but also to give calm and relaxation. In this sense, it manages to change one of the main characteristics of architecture, namely its immobility, which, thanks to the presence of the liquid element, is vivified with a potential mutability.

博登公园自然游泳池_Borden Park Natural Swimming Pool/gh3* architecture

水上公园_Water Park Aqualagon/Ferrier Marchetti Studio

Termalija家庭健康中心_Termalija Family Wellness/ENOTA

"如水一样，我的朋友"_"Be Water, my friend"/Diego Terna

my friend"

扭转间开辟出巨大的空间来容纳水流，或是再一次展现多样的形式和高端的技术系统，来将水元素融入一系列发光装置内。

水，它是游戏，是强大的力量，能带给我们惊喜，使建筑师不断更新作品，设计出光彩熠熠的建筑，准确地说是利用水本身进行创作：用水做成冻结的瀑布、发光的空间构造，将水流声化作流动的节奏韵律。这些建筑作品实现了李小龙最著名的一句名言：如水一样，我的朋友。

The examples that will be presented here reflect on the relationship between water and architecture, deforming and twisting themselves because of it, opening huge cavities to accommodate it or, again, defining formal and technological systems to incorporate it into a series of luminous paths.
Water, as game, brute force, surprise, allows us to build shimmering works, renewed continuously, precisely by exploiting the water itself: they become frozen waterfalls, luminous caves, rhythmic accompaniments of flows; they fulfill one of Bruce Lee's most famous suggestions: *Be water, my friend.*

"如水一样，我的朋友"
"Be Water, my friend"

Diego Terna

水教会人类
不要阻碍水平延伸。
让我们倾听光明
不要阻挡垂直倾泻。

——爱德华多·奇利达，作品集（2005年）

水平/垂直

2008年，正值宫崎骏执导的电影《波妞》的巅峰时期，这部动画电影中的两位主人公一早醒来，发现自己家这栋位于山顶的房子，竟是昨晚可怕海啸来袭后的唯一幸存居所。

电影全程的选景是连续起伏的悬崖和山丘，而海啸过后这片景观完全平整了：一张水平的桌子静静安放在地面上，挡住了海啸造成的破坏和剧变。

电影中的水呈现给我们一种双重角色，除海啸带来的破坏之外，它也是快乐力量的集中，鲜活如有生命，吸收了波妞的精髓而充满活力。波妞原本是海洋中的"动物"，但却选择成为人类小女孩与人类相爱，她能摆脱来自水本身及海洋环境的压力。

El agua enseñó al hombre *The water taught man*
A no bloquear la horizontal. *Not to block the horizontality.*
Escuchemos a la luz para *Let's listen to the light*
No bloquear la verticalidad. *Not to block the verticality.*

– Eduardo Chillida, Escritos (2005)

Horizontal / Vertical

At the peak of the movie Ponyo, directed by Hayao Miyazaki in 2008, the two protagonists of the animated feature film wake up in their home at the top of the hill, discovering that they are in the only building not flooded by the terrible tsunami of the previous evening.

The landscape, which during the course of the film we discovered was a continuous up and down of cliffs and hills, is now completely flat, homogeneous: a horizontal, calm table, which totally hides the violence and upheaval caused by the tsunami.

In this dual role, the water presents itself in the movie as a joyful concentrate of power, an animated being, made vigorous by the very essence of Ponyo, "animal" of the sea but child by choice, in love with humans and capable to get rid of the pressure of the water itself, its natural environment.

The house, on the other hand, assumes the role of the vertical counterpoint, as if to speak for every form of architecture that defined the city, which now lies under the sea: it is the element that tears the homogeneity, the horizontal line, constructing the memory of the space built by humanity.

Water can be destruction.
Water can be life.
Water can be the mirror of the actions that take place in its vicinity.

另一方面，房子扮演着与水垂直对立的角色，似乎在为那些已沉入海底的、代表着这座城市的每一种建筑形式诉说着什么。作为一种打破同质结构和水平线的元素，它构建了人类打造的空间的记忆。

水可以破坏生命。
水可以养育生命。
水是其周围一切行为的镜子。

从这个意义上来说，正如我们在电影中看到的，建筑在水出现后找到了自己存在的价值，从周围的环境中脱颖而出：正如水在不断运动一样，建筑一直保持静止。但恰恰在这两种现实的比较中，建筑获得了一种更闪亮的特征，这种特征也许更快乐、更有趣。另一方面，正是因为水没有形，才突出了建筑的形。

如水一样，我的朋友
"像水一样穿过裂缝。凡事不要过于绝对化，要适当调整你的目标，你会找到一条迂回前行的路。如果你内心没有任何东西是坚定不变的，外在的东西就会让你暴露无遗。保持空灵之心，无式，无形，就像水一样。水倒入杯中就成了杯子的形状，倒入瓶中就成为瓶子的形状，倒入茶壶中就成为茶壶的形状。而水能载舟，亦能覆舟。像水一样吧，我的朋友。"

——李小龙，皮埃尔·伯尔顿展访谈（1971年）

也许因为水之无形，以及它适应任何形状的能力，它成为建筑空间的理想伴侣；它填补了建筑的空缺，使建筑的意义更加显著，展现了建筑之美。正如李小龙话中所示，水秉持坚韧不屈的灵活态度，它能摧毁周围的环境，但在某

In this sense, as we see in the film, architecture finds in the presence of water a reason for development, to emerge with respect to the surrounding area: as much as the water is in constant movement, so then is the architecture immobile. But precisely in the comparison between the two realities, architecture acquires a more shimmering character, perhaps also more joyful and playful. On the other hand, water, which has no form, brutally highlights the very form of architecture.

Be Water, my friend
"Be like water making its way through cracks. Do not be assertive, but adjust to the object, and you shall find a way around or through it. If nothing within you stays rigid, outward things will disclose themselves. Empty your mind, be formless, shapeless, like water. If you put water into a cup, it becomes the cup. You put water into a bottle and it becomes the bottle. You put it in a teapot, it becomes the teapot. Now, water can flow or it can crash. Be water, my friend."

– Bruce Lee, Interview at the Pierre Berton Show (1971)

Perhaps this formless being of water, and therefore its capacity to adapt to any shape, singles it out as the ideal partner of built space: it fills its voids, increases its meanings, brings out its beauty. As Bruce Lee implied, the attitude of water is a form of relentless flexibility, capable of destroying its surroundings, but, in a certain way, also of enhancing it. Again, water brings an element that architecture, in general, does not possess: its being formless, flexible, changeable, tends to underline, but above all to lighten, the static, the rigidity, the extreme search for defined forms typical of architecture.
Goethe, defining architecture as frozen music, referred to an analogy between the liquid, with its characteristics of movement and fluidity, and the frozen object, which reaffirms, in a rigid form, precisely the potential of movement and fluidity. The combination of water and architecture helps us to reflect

种程度上讲，也能渲染周围的环境。同样，水带来了建筑通常所不具备的元素：水之无形、灵活、可变，倾向于强调建筑形式，但首先要削弱建筑的静态、僵化等特征，以及减少其对典型形式的极端追求。

歌德将建筑定义为凝固的音乐，指的是具有流动特征的液体与静止物体之间的协奏曲，这种严格的类比再次说明，建筑中可以融入动态和流动元素。水和建筑的融合让我们思考元素相融的可能性，思考如何让建筑空间从静止状态中解放出来，以容纳功能设施和行为艺术。

另一方面，就像在电影《波妞》中我们理解的那样，水也是一种游戏，它设法淡化了建筑中最有层次的部分，允许前所未有的视觉体验、意想不到的思考和富于联想的反馈；简而言之，元素的融合赋予建筑空间本身所不具备的可变性。

于水上漫步

因此，建筑与水的关系似乎是产生于彼此的对立状态：垂直对水平，刚性对柔性，不变对可变。但是水也能在奇妙的自然律动中给我们带来梦幻体验，这种体验是精神和心理层面的———它将我们带进一个光明的世界，在那里我们的身体失重、体积飘忽不定、形体变幻万千；在那里我们漂浮遨游，种种感受宛如在天空飞翔。

2016年，在意大利的伊索湖上，两位保加利亚裔美国艺术家克里斯托和珍妮-克劳德利用漂浮码头成功实现了这一梦想，但不是飞行，而是让游客在水上漫步。

他们为实现这个想法而在水面上建造了一个平台，平台连接着湖岸和湖中央的一座岛屿，创造了一个前所未有的惊奇宇宙。这是人类第一次在没有交通工具的情况下，只身穿过水域。

随着湖水的轻微波动，一条橙色的织物如呼吸般起伏，在景观和游客之间，在人行小道和令人兴奋的水元素构

on this potential, on how the built space could be if freed from its need to stand still, in order to accommodate functions and actions.

On the other hand, water is also a game, as can be understood in Ponyo, and as such it manages to play down the most hieratic part of architecture, allowing unprecedented visions, unexpected reflections, suggestive reverberations, in short, giving that quality of changeability that the built space does not possess.

Walking on Water

The relationship with water, therefore, seems to emerge by opposition: vertical/horizontal, rigid/flexible, immobile/changeable. But water also has the ability to bring us into the bed of a dream, in a wonderful physical fluctuation, which becomes mental and psychological – it leads us into a light world, where our body loses weight, bulk, shape. It is the world of swimming, of floating, which makes us imagine, conversely, that we are flying.

In 2016, on Iseo Lake, Italy, Christo and Jeanne-Claude manage to transform this dream into a reality, with the Floating Piers: not flying, but walking, the Bulgarian-American artists allow visitors to literally walk on the water.

They do this by building a platform lying on the surface of the water, connected to the shore and to an island in the middle of the lake, thus creating an unprecedented universe of disconcerting surprise. For the first time it is possible to cross the waters with one's body, without the intermediary of a vehicle.

Through a strip of orange fabric, which rises and falls as if breathing, following the slight wave motion of the lake, a new relationship is established between landscape and people, between the slow temporality of the physical path and the exciting size of the body of water: visitors are transported to a usually inaccessible place, thus being able to enjoy an unprecedented point of view; this permits them to grasp the surrounding environment in a never experienced form, feeling the hills surrounding the lake as new life companions, living beings, able to interact with the tiny dimensions of other people, standing in the middle of the void.

Here, water is the means to outline a situation that lasted, like a dream, only a few days: it builds support

造之间，建立了一种新的联系：游客由此可以到达之前所无法到达的地方，从而能够以前所未有的视角观赏别样的景观；这使得他们以一种从未体验过的方式来感受周围的环境，感受到湖周围的山成为自己新的人生伴侣，也使站在水上平台上的游客能够与他人的微小维度进行互动。

在这里，水勾勒出一场如梦般的只能持续几天的情景：它为游客活动提供支撑，放大人类的能力，无限增强快乐和趣味性。水和游客一起呼吸，成为游客身体的一部分，人和湖水同呼吸、共命运。

水成为探索发现的场所，成为思考和想法的元素：伴随着心脏跳动，人们走在平台上，体验着那种烂漫无邪的情感，做一些平日里被禁止的、超自然的事情。从某种意义上来说，水颠覆了物质世界的规则，勾勒出身体想表达的行动自由，这提供给我们一种新的平衡。

建筑/水

下面我们要看到的建筑项目都将水作为其结构开发的关键元素，作为其外部几何形状的基底。这主要是通过对立关系来完成的，建设虚拟的地形景观，使建筑整体保持垂直、高高耸立，而水则与之相反，呈现水平状态。

费里尔·马切蒂工作室的水上公园项目（210页）描绘了一种山地地形，通过一系列镂刻在一起、很像一种几何螺旋形式的破碎平面，借助抽象形式呈现精确形态。因为水是流动的，所以看似由不规则部分相交而组成的建筑看起来很紧凑，但是融入水元素后，建筑仿佛置身山地地形中。

于此，歌德的定义似乎比在其他情况下都显得更为真实：凝固的音乐，或者更准确地说，凝固的液体。我们对面是一座冰山，它像雕塑一样从水中升起。或许，它可能是个瀑布，在下降的那一刻被冻结凝固在空中一动不动，但其下冲的气势完好无损。

for human actions, it amplifies their caliber, it increases their joyful and playful outreach. It breathes with them, becoming part of the physique, allowing a total symbiosis between people and the lake.
Water becomes the place of discovery, the element that allows reflections, thoughts: with a skipped heartbeat, people walk on the platform, experiencing the almost childish emotion of doing something prohibited, supernatural. In a sense, water subverts the rules of the physical world and outlines a moving freedom of body expression, which gives us a new balance.

Architecture / Water
The projects we will see below make water the key element of their structural development, a place around which to base their shape, their geometry. They do it primarily by opposition, building fictitious topographies that allow a remarkable elevation, outlining the forms of verticality, in contrast, or dialogue, with the horizontality of the water.
Ferrier Marchetti Studio's Water Park Aqualagon (p.210) defines a sort of mountainous territory, rendered abstract, but precise in its conformation, by a series of broken planes that chase one another, like a sort of geometric spiral. Insofar as the water is fluid, so too the building is nervous, made of segments that seem to intersect without a rule, but which then find themselves in an almost mountainous topography. Here, more than in other cases, Goethe's definition seems to come true: frozen music, or more accurately, frozen liquid. We are faced with a sort of iceberg, which seems to be born from the water itself, rising like a sculpture. Or, perhaps, it is a waterfall, frozen at the very moment of its peremptory descent, remaining motionless, with its potential intact.
Even the Termalija Family Wellness (p.222) by ENOTA works by creating a new topography, which, however, does not turn into concretion of water, but rather in a covering: it builds a cave, a protective cover of the internal environment, capable of creating a light, sunny, aquatic space, despite the alleged heaviness of the cover. The interior and exterior work on different levels: the external part becomes a sign of architecture, it shows itself in the surroundings, as if to underline the presence of water. It does this through a domestic, urban reference, dividing the wide roof span into a series of fractal subsets, reminiscent of the roofs of a traditional village. Inside, the covers are an innervated texture of light, a series

ENOTA设计的Termalija家庭健康中心(222页)也是通过创造这样一种新的地形来实现的,然而,这种地形并非是将水凝结,而是建造了一个保护罩:用覆盖物来建造洞穴,形成内部环境的保护罩,从而创造一个阳光充足的水上空间,尽管据说这个保护罩很重。结构分为内外两层。外层成为建筑的标志,它在周围环境中展示自己,似乎在强调水的存在。这一构造是通过参考国内其他城市做到的,将宽阔的屋顶划分成一系列不规则单元,让人联想起传统村庄的屋顶。而内层的保护罩是一种感光材质,一系列形状断断续续的圆顶相互交织,那形状让人联想起海浪,在光线的追逐下,水和其反射的光影欢快对话。

由gh3*建筑公司设计的博登公园自然游泳池(198页)将水和建筑的特征结合在一起,两种元素并行不悖,形状并不是变幻多端,但是却更好地烘托了平静的氛围,仿佛在等待未来的巨大改变。

这里看起来像是一个悬浮的世界,几乎就像克里斯托在伊索湖建造的作品试图定义的感觉一样:不需要展示威力,建筑师在水面上建造了一系列通道,伴随着游客感受水本身的轻盈;这些通道旨在营造一种轻松的环境,指引着游客轻轻漂浮在水面之上。

内部/外部

在处理建筑与水的关系时,建筑师用独特的方式,使所选项目的外部和内部之间构成连续的比较,这似乎是将液体元素融入建筑物的共同特征。

正如李小龙在20世纪70年代所描述的那样,每一座建筑都试图融入水这一元素,将它无形但却十分强大的力量加以利用。这种对水的力量的驾驭似乎在阿夸拉贡水上公园中体现得十分明显,建筑师甚至扭曲了这座水上公园,以保持对水的坚定立场。虽然公园不能完全做到掌控水的力量,但是建筑师还在不断尝试中,正如波妞在海浪上奔逐那样:这种力量很残酷,但却带着孩童般的乐趣。

Termalija家庭健康中心也是如此,其内部的保护性元素被封闭在内部空间:水是一种珍贵的必需品,必须加以

of broken domes, twisted in a shape that recalls the waves of the sea, a continuous chasing of lights, which thicken the dialogue of reflections with the water below.

The Borden Park Natural Swimming Pool (p.198) by gh3* architecture works instead on a combination of characters, that of water and that of architecture, carried on similar tracks, in which the forms tend not to be exuberant, but, better – measured, calm, as if waiting for a future upheaval.

It looks like a world in suspension, which almost seems to refer to the same feelings that Christo's work on Iseo Lake was trying to define: without the need to show muscles, the architects define a series of passages on the water which accompany visitors along the lightness of the water itself; these passages attempt to provide that condition of levity that leads people to float lightly above the water itself.

Inside / Outside

In their peculiar way of dealing with the relationship with water, the selected projects construct a continuous comparison between outside and inside, which seems to be a common feature of approaching the liquid element.

Each of the buildings attempts to compress the power of water, trying to harness its being formless, yet so powerful, as Bruce Lee described in the 1970s. This harness seems manifest in Aqualagon Water Park, which almost twists on itself to be able to maintain a firm position towards water. It cannot completely do it, but in trying it seems to play with the water itself, as Ponyo did with the waves of the sea: there is power, a brutal force, but channeled into an almost childlike sense of fun.

So also does Termalija Family Wellness, with its protective element towards the interior enclosed in its space: water is a precious necessity, which must be preserved and for this reason the relationship between outside and inside is filtered by architecture. There is therefore no moment in which the aquatic environment is exposed to the outside: light, air and heat are brought inside thanks to the mediation of the forms of architecture themselves.

In the Borden Park Natural Swimming Pool, a different relationship is established, in a certain sense less playful – more formal, as if it were a didactic element; the architectural complex behaves like a teacher,

保存，于是建筑师通过技术手段，将室外环境条件纳入室内。因此，水环境时刻保存在室内，光、空气和热量全部融入室内，这要归功于建筑形式本身的调和。

在博登公园自然游泳池项目中，建筑师在建筑物与水之间建立了一种不同的联系，在某种意义上，这种联系少了几分玩味，多了几分正式，好像是一种学究式的元素，整个建筑综合体呈现出教师的气质，将水融入，并使其穿过一系列空间得以净化。这是一个连续的循环，需要特定装置不间断运行，也许正是因为这个原因，建筑变得更庄严、更正式，以便捕捉液体元素的丰富形式，并将水引向其一般不会流到的地方。建筑因此成为一种示范行为，在这个意义上，它严格遵循"水平"的节奏，这是一种音乐伴奏，它打破了水流和游客的时间限制。所有的建筑都倾向于尽可能地向周围环境敞开内部空间，就好像在处理水的时候，有必要兼顾整个区域一样。我们在宫崎骏的电影中看到的山上房子是海啸中唯一幸存下来的建筑，回到我们分析的项目所定义的维度：在这部电影中，水隐藏了这个地方的记忆，而幸存的小房子成了唯一记忆清晰的地方，最终它迫使大海自己撤退，陆上人们的生活恢复如初。

在这里，水并不是一种令人恐惧的元素，但它本身却带来了一个地区的记忆、一个净化的过程：建筑的任务是把自身置于整个建筑结构的边缘，面对水的流动，波澜不惊。

于是，它们着手于接受这种追寻记忆的工作，带游客去到一个更梦幻的地方，在那里他们可以享受水的温和力量。这是在伊索湖上发生的事情，沿着克里斯托的平台行走，随着波浪运动摆来摆去：你可以感受到水的力量，你可以感受到波浪运动的势不可挡；然而，快乐是并存的，那股波浪的力量去迎合、去依偎，与走在流动水上道路的游客并肩齐行。

也许在这里，在本文介绍的项目中，可以看到这种普遍的态度：渴望用游戏、用幸福、用意想不到的快乐给人们安慰。水，即使以其最平常的形式展现，也绝不平庸，正如本文讨论的三个案例那样，只要在明确界定的范围内加以利用，就能够营造这些感觉。

如水一般吧，我的朋友，或者更理想状态，水，做我的朋友吧。

incorporating the water and forcing it to go through a series of spaces to be purified. It is a continuous cycle, which requires a tireless movement: perhaps for this reason architecture becomes more solemn, more formal, in order to capture the exuberance of the liquid element and to channel it into places from which it would tend to escape. Architecture thus becomes a demonstrative act and, in this sense, follows very "horizontal" rhythms, a sort of musical accompaniment, which beats the time of flows and visitors. All buildings tend to open their internal void as much as possible towards the surrounding environment, as if it were necessary to refer to the territory, when dealing with water.

The image of the house on the hill, the only built bastion that survived the tsunami, which we saw in the Miyazaki movie, returns in the universe defined by the analyzed projects: in the film, water hid the memories of the place and architecture became the only point in which the memory was explicit, forcing the sea itself to withdraw, to return to a daily normality.

Here water is not such a frightening element, but it brings with itself the memory of a territory, a community, a purification process: the task of architecture is to place itself at the edge of those systems, facing fairly the fluid movements of the water.

Thus, they proceed in a work of acceptance of this memory, taking visitors to a more dreamy landscape, where they can enjoy, in a benevolent form, the power of water.

This is what happened on Iseo Lake, walking along Christo's platform, tossed by the wave motion: you could feel the power of the water, you could feel the possible annihilation given by the wave motion; yet there was joy, as if that same power had bent to welcome, to cradle, to become small and minute like the people who walked the liquid path.

Perhaps here, in the selected projects, this general attitude can be seen: the desire to give comfort to people, with the game, with the well-being, with the joy of the unexpected. Water, which is never banal even in its most "normal" forms, is capable of building these sensations, provided that it is channeled within well-defined limits, as occurs in the three cases discussed.

Be water, my friend, or better, be my friend, water.

博登公园自然游泳池
Borden Park Natural Swimming Pool

gh3* architecture

加拿大城市公园中的这座游泳池由石笼墙围成，内置生态过滤装置
A pool with a gabion walled rectilinear building in the Canadian civic park involves an eco-filtration process

博登公园自然游泳池是加拿大第一个不含化学物质的公共室外游泳池。该项目取代了原有泳池，为这里的400名游泳者建造了应季亭和景观区。建造这座泳池时，建筑师遇到的挑战是需要为这个规模较大的水池配备高质量的水控装置(这是建造公共游泳设施的基本标准)，同时达到环境健康标准，实现自然过滤过程。设计师首先开发了一个泳池技术，通过石头、砾石、沙子和植物过滤原理来净化水，这启发了为更衣室设施创造一个以材料为导向的设计理念，实现了严格的技术控制和美学一体化设计。

加拿大公共泳池的建造要求是世界上最严格的，因此为了实现这个项目，建筑师需要采取创造性的方法，将这个项目归类为"娱乐水域"，建筑许可证对该项目的定义为"差异并存的建筑海滩"——差异体现在游泳池上。泳池包括一个平衡的生态系统，植物、微生物和营养物质经过砾石和沙子的过滤集装在一起，形成"活水"，没有用到土料。过滤有两种方式：一种是通过生物-机械系统过滤，另一种是借助人工湿地、砾石过滤器和原位浮游动物进行过滤。这是一种不加消毒液、不使用化学物质的过滤系统，其中的隔离膜在循环过程中会用到水，并通过自然过程进行清洗。这种过滤方式被用在游泳池的北端。

在平台上，水流过沙石浸没池和一个种着水生植物的池子。在这些池子附近，是一个与建筑相连的由石笼墙围住的颗粒过滤器PO_4(磷酸盐)吸附结构。

该建筑包括通用更衣室、淋浴、盥洗室、工作区和水过滤装置。游泳池项目包括一个儿童游泳池、一个深水池、平台上的室外淋浴间、沙滩、野餐区和其他与游泳池相关的娱乐活动空间。

该项目所使用的材料在泳池的技术需求与建筑围护结构和景观要素设计之间建立了基本的概念联系。石笼墙结构的深色石灰石和钢质构造，将外立面的垂直面打造为过滤器般透气的粒状多孔结构。一片平面景观将游泳池划分为两个区域，这片景观平台将沙滩、混凝土泳池周壁和木质平台无缝衔接起来。低矮直线型建筑的石笼墙以一个像盖子一样的平屋顶为收尾点，这个屋顶框柱了远处公园里树木的树冠景色，增强了泳池区内开阔的空间感。该项目通过选取独特的元素形式、降低材料损耗，使用户体验轻松愉悦，在景观中淋浴也成为一种享受。不同建筑元素的并列让人联想到北萨斯喀彻温河的地质和草原边缘的平坦地形。

该建筑借鉴了两个相邻的中世纪现代风格的泳池建筑。这些建于

20世纪50年代的建筑勾勒出泳池区西南边缘的轮廓，并将新建的泳池与一个有着百年历史的独特建筑文化遗产——市民户外浴场——连接在一起。

博登公园自然游泳池象征着埃德蒙顿市的模范领导作用，表现了对公共基础设施建设中优秀建筑师的认可。自1924年以来，在过去将近一个世纪的时间里，博登公园已经发展成为一个共享的户外娱乐活动场所，是多数人家庭聚会和户外洗浴的最佳选择，对当地社区乃至埃德蒙顿市具有巨大的民生价值和社会意义。而该项目成为公园历史景观不断发展变化的地标性元素。

The Borden Park Natural Swimming Pool is the first chemical-free public outdoor pool to be built in Canada. The project replaced an existing pool, with a seasonal pavilion and landscaped precinct for 400 swimmers. The architectural challenge was to create a largescale pool with high quality water control (an essential criterion for any public bathing facility), while also achieving an environmentally healthy and natural filtration process. The designers began by establishing a pool technology that cleanses the water through stone, gravel, sand, and botanic filtering processes, which inspired the creation of a materials-oriented concept for the changeroom facility, to achieve a technically rigorous and aesthetically integrated design. Canada's guidelines for public pools are some of the strictest in the world, so to realize the project the architects needed to take a creative approach, classifying the project as "recreational waters" with the building permit as a "constructed beach with variances" – the variances being the pools. The pool involves a balanced eco-system where plant materials, micro-organisms, and nutrients come together within a gravel and sand filtering process to create "living water"; there is no soil involved. Filtration is achieved in two ways: by means of a biological-mechanical system or the constructed wetland and gravel filter, and in situ, with zooplankton. This is an unsterilized, chemical and disinfectant free filtering system in which isolating membranes contain water as it circulates and is cleansed via

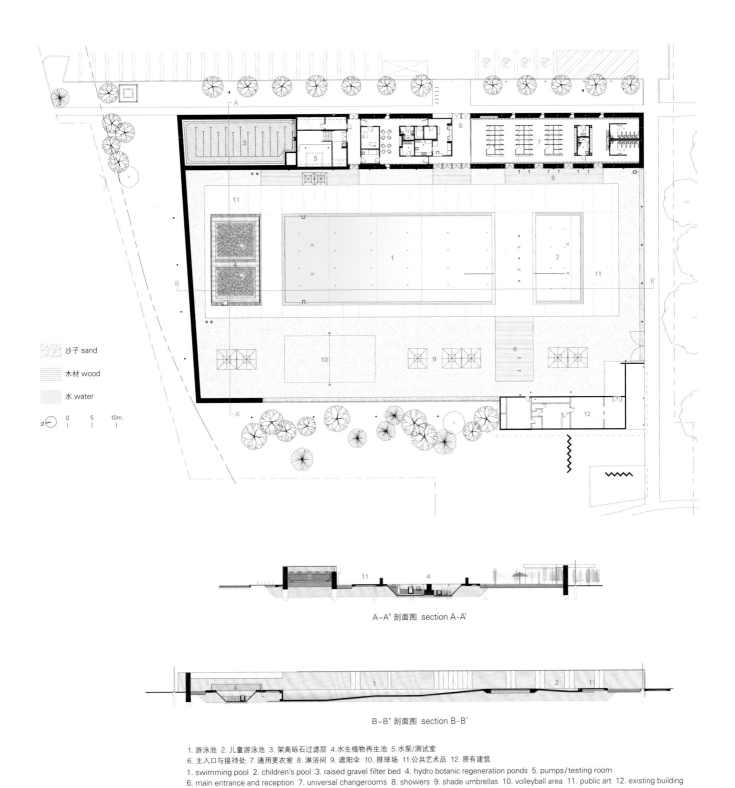

项目名称：Borden Park Natural Swimming Pool / 地点：Edmonton, Alberta, Canada / 建筑、景观、室内设计：gh3* architecture / 项目团队：Pat Hanson, Raymond Chow, Bernard Jin, Joel DiGiacomo Nicholas Callies, John McKenna, DaeHee Kim / 结构、机械、电气、土木、LEED工程：Morrison Hershfield / 石笼墙工程：Associated Engineering 自然游泳池工程：Polyplan GmBh / 总承包商：EllisDon / 客户：City of Edmonton contact Carol Belanger City Architect / 所有者：City of Edmonton / 总建筑造价：$14.4M CAD 总施工成本：$11.2M CAD / 用地面积：5,989.5m² / 建筑面积：770m² / 竣工时间：2018.7 / 摄影师：©Raymond Chow (courtesy of the architect)

a natural process. This process takes place at the north end of the pool precinct. On deck, water passes through a sand and stone submersive pond and a planted hydro botanic pond. Adjacent to these ponds, a granular filter PO_4 adsorption unit is enclosed by the gabion walls continuous with the building. The building houses universal change rooms, showers, washrooms, staff areas and the water filtration mechanisms. The swimming program includes a children's pool, a deep pool, on-deck outdoor showers, a sandy beach, picnic areas, and spaces for other pool related recreational activities.

The project's materiality creates a fundamental conceptual connection between the technical demands of the pool and the design of the built enclosure and landscape elements. The dark limestone and steel of the gabion wall construction defines the enclosure's vertical dimension as filter-like or breathable, as granular and porous. The pool precinct is defined by a planar landscape where flush to surface detailing creates seamless interfaces among sandy beach, the concrete pool perimeter and wood decking. The gabion walls of the low rectilinear building terminate with a lid-like flat roof that frames the tree canopy of the park beyond and enhances the sensation of open-sky spaciousness within the pool precinct. The elemental form and reductive materials ease the user experience and enrich the narrative of bathing in the landscape. The juxtaposition of the constructed elements invokes comparisons with the geology of the North Saskatchewan River and the flat topography of the prairie lands edge.

The building takes a cue from the two adjacent mid-century modernist style pool buildings. Built in the 1950s, these extant buildings define the southwest edge of the pool precinct and connect the new with a century-old cultural heritage of architecturally distinctive, civic outdoor bathing pavilions.

The Borden Park Natural Swimming Pool is indicative of the City of Edmonton's exemplary leadership and recognition of the civic importance of architectural excellence in the building of public infrastructure. Borden Park has evolved over the last century as a place of shared outdoor recreational activities, as a destination for family gatherings and outdoor bathing since 1924. It holds immense civic value and social significance for the communities and for Edmonton. The project contributes a landmark element within the ongoing transformations of the historic landscape of the park.

a 游泳池
b 儿童游泳池
c 架高砾石过滤层
d 水生植物再生池
e 水泵/测试室

a swimming pool
b kid's pool
c raised gravel filter bed
d hydro botanic regeneration ponds
e pumps/testing room

3
经过过滤的水因为重力流入水生植物再生池

3
filtered water gravity flow to hydro botanic regeneration ponds

2
混合"灰"水被水泵抽到架高砾石层处

2
mixed "grey" water pumped to raised gravel bed

1
"灰"水从游泳池周边的排水槽溢流

1
"grey" water overflow from pool perimeter gutters

4
经过完全过滤和充气的清洁水流入测试室

4
fully filtered and aerated clean water flow to testing room

5
清洁的泳池水被水泵抽到泳池底的供水管口处

5
clean swimming pool water pumped to supply nozzles at base of pools

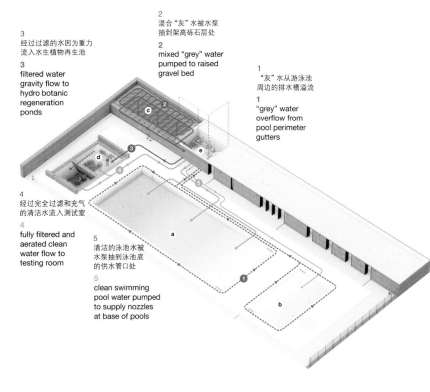

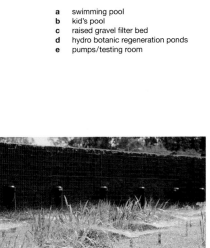

水上公园
Water Park Aqualagon

Ferrier Marchetti Studio

巴黎附近一座拥有透明的几何穹顶及穹顶上蜿蜒的人行小径的水上公园
A water park near Paris defined by transparent and geometric dome with walkable winding paths

这个项目的平面设计灵感来源于风向和太阳的轨迹。水上公园坐东朝西，背靠森林，可以免受冬季寒冷东北季风的影响，在温暖的天气可以充分利用凉爽的微风散热，而西南朝向可以让公园全年都沐浴在阳光中。

如同折纸雕塑一般的水族公园景观建筑有35m高，直指天空。从周围的区域望去，该结构清晰可见，可作为参照点，与相对平坦的地形形成对比。然而，它又不是强加在景观上的元素，而是景观本身的自然延伸。水上公园由大片水域所围绕，似乎是从湖中浮现出来的——一个由瀑布、薄雾、蒸汽和水生植物组成的世界，与水和透明的建筑元素交相辉映。

水上公园室内空间的上方有一个绝佳的观景点，在这里可以欣赏到广阔的景致。露台向行人开放，一条通道环绕建筑盘旋而上，使木板人行道一直延伸到顶端。游客也可以乘坐电梯来到人行道顶端欣赏风景。

随着"折纸"展开，更大的空间呈现眼前，大玻璃板的设计使自然光可以进入室内，给游客一种内外连通的感觉。大楼顶部的透明几何穹顶让游客在游泳的同时可以眺望天空。穹顶形成整个水上公园的建筑轮廓。结构线条和屋顶轮廓经过设计优化，从人们的视野中消失，独留天空浮现眼前。

该项目的所有元素都营造出一种景观感，有助于提升公园中水上活动的体验效果。种有植物的露台所创造出的生态系统颇有环保教育意义。与环境可持续发展相关的各种元素——地热能源和水循环——结合在一起，形成了一套连贯的主题来构建整个项目，其环保内涵一目了然。

该项目的理念是将水上公园层层叠加，并延伸到日光休息室、露台和餐厅中。这些区域的活动人流量会随着季节变化而变化。带礼堂的游艇会在晚上或者一天中的不同时间来到，为观众提供音乐会和表演场地。

水上公园的入口与一个大型前庭相连，前庭周围环绕着自然村的不同公共空间，如论坛和体育中心。从面向湖边的大厅，经过一条长长的走廊可以到达更衣室以及机房、储藏室等附属空间。

游客一进入更衣室区域，就能见到水上公园内部，这里由一大片水域构成，散布着各种大小的岛屿，不同的岛屿功能各异。几座小岛延伸到外面的环湖中，从而强调了室内和室外的连续性。

The direction of the winds and the path of the sun determine the floor plan of this project. Protected from cold north-easterly winter winds, nestling up to the forest, the aquatic park opens towards the west to make the most of cool breezes in warm weather, whilst the south-westerly aspect bathes the park in light throughout the year.

Like a sculpture of origami, the proposal for the aquatic park resembles an unfolding landscape of 35m height as it rises into the sky. The structure is clearly visible from the surrounding area, becoming a point of reference, contrasting with the relatively flat topography. Yet it is not an element which has been imposed on the landscape, but an extension of the landscape itself. Located by a large expanse of water, the aquatic park appears to emerge from the lake – a world composed of waterfalls, mist, steam and aquatic plants, which plays with water and transparency.

The structure presents a stunning vantage point over the inside of the aquatic park and offers spectacular views the wider landscape. The terraces are open to walkers; a circuit offers a walk surround the building, extending the board walk promenade. A lift offers the occasion to climb to the top of the walk

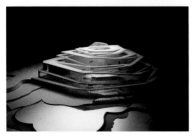

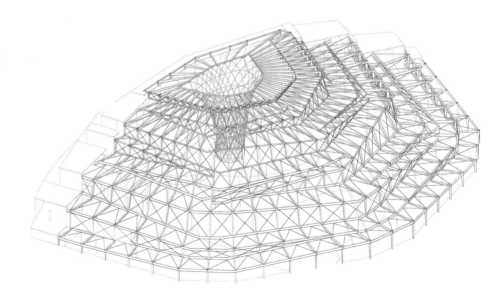

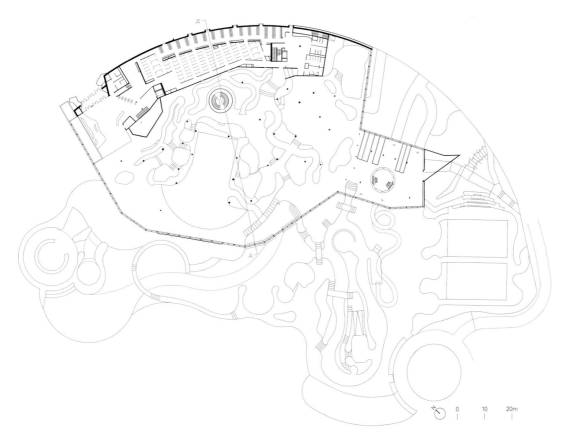

to enjoy the view.

As the origami unfolds, it creates empty spaces, which are filled by large glass panels that let natural light inside, and give visitors the sensation of a continuity between the interior and exterior. At the top of the building, a transparent geometric dome allows visitors to swim while watching the sky. This dome forms the silhouette of the aquatic park.

The structure and the roofing have been optimized to disappear in the sight lines from the basins and accentuate the presence of the sky.

All the elements of the project participate in creating a sense of spectacle which serves to heighten the experience of the aquatic activities in the park. The remarkable ecosystem established by the planted terraces forms an instructive path. The various elements related to environmental sustainability – geothermal energy and water recycling – come together to form a coherent narrative that structures the entire project, and can be clearly read by the public.

The principle of the project, with its accumulation of levels, extends the aquatic park into sun-lounges, terraces and restaurants. These activities can easily be linked to seasonal variations in attendance. An auditorium barge arrives in the evenings and at various times of day to offer concerts and shows.

The entrance to the aquatic park is connected to a large forecourt around which Villages Nature's different public spaces are arranged, such as the forum and sports center.
The hall, which opens onto the lake, leads to changing rooms via a long corridor and ancillary spaces including machine room, storage.

Once visitors have gone through the changing room area, they discover the inner space of the aquatic park, designed as a large expanse of water strewn with islands of various sizes and with a variety of functions. This archipelago extends outside, into the lagoon, accentuating the continuity between indoors and outdoors.

项目名称：Water Park Aqualagon / 地点：Villages Nature Paris, next to Marne-la-Vallée (77), France / 事务所：Ferrier Marchetti Studio / 外围护结构与结构工程：C&E Ingénierie, with Henry Bardsley et Wolfgang Winter Ingénieur consultants / 景观：Interscene Thierry Huau / 救生站设计：Sensual City Studio / 高环境质量助理：Transsolar
流畅设计：Inex / 声学设计：Peutz / 照明：Atelier Audibert / 认证：HQE certification (High Quality Environmental standard) sport facilities - swimming pool V2 Exceptional level
客户：Villages Nature Paris (Pierre & Vacances-Center Parcs and Euro Disney SCA) / 功能：aquatic complex / 面积：8,000m² / 竣工时间：2017 / 摄影师：©Luc Boegly (courtesy of the architect) - p.210~211, p.214~215, p.218, p.219; ©Didier Boy De La Tour (courtesy of the architect) - p.213, p.220~221; ©Hugo Deniau (courtesy of the architect) - p.212

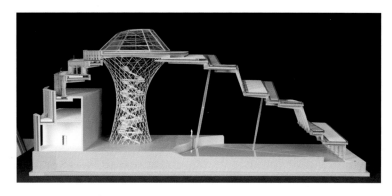

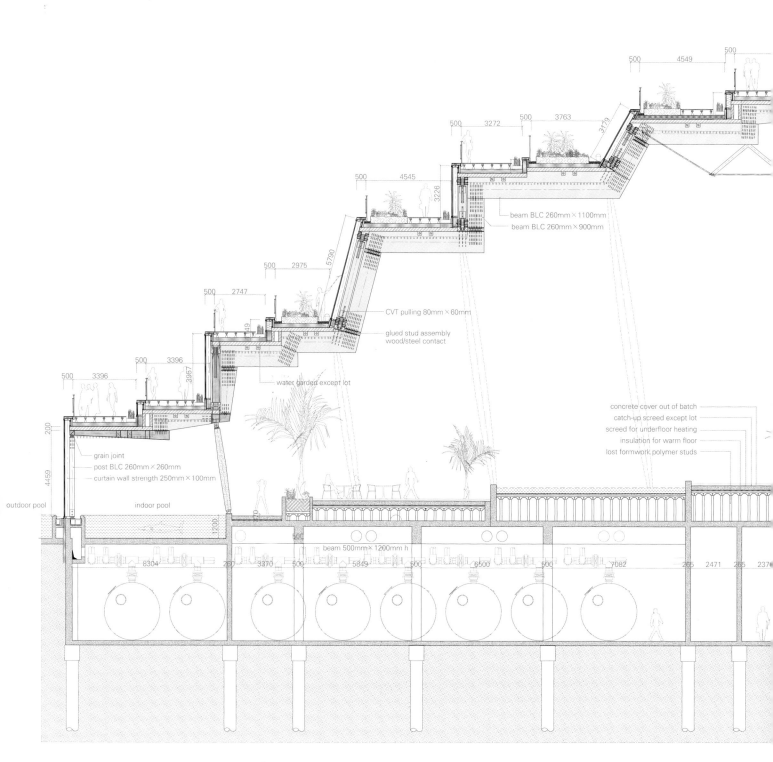

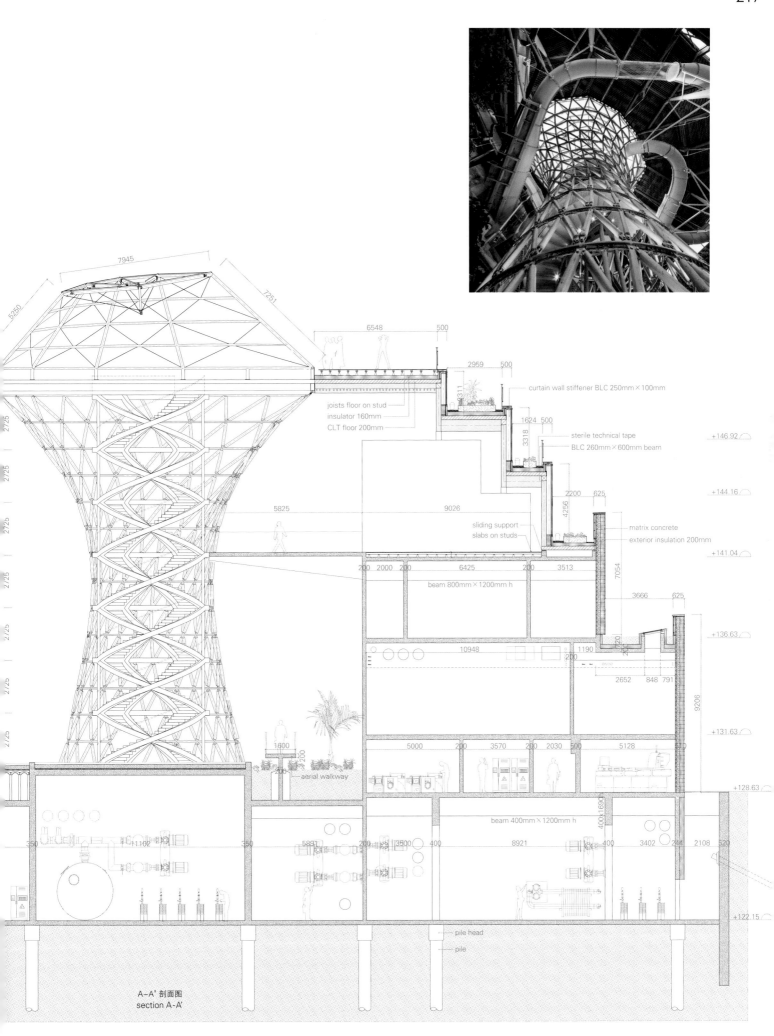

A-A' 剖面图
section A-A'

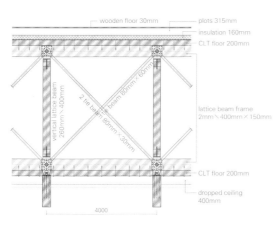

网格梁标准立面
standard elevation on lattice beam

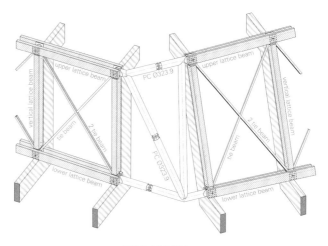

网格梁连接处轴测图
axonometry of lattice beam junctions

Termalija家庭健康中心
Termalija Family Wellness
ENOTA

覆盖斯洛文尼亚健康中心的四面体屋顶与周围环境相协调
Tetrahedral volumes of roof covering Slovenian wellness complex are harmonized with the surroundings

过去15年间，ENOTA在Termalija建造了一系列项目，而Termalija家庭健康中心是其中最新的一个，它实现了从20世纪80年代的经典健康中心到现代化温泉疗养中心的转变。

新疗养设施所处的位置以前是用来放置泳池遮盖物的，冬季时用来覆盖室外游泳池。夏天时会把遮盖物收起来，为稀少的户外空间腾地方，但实际情况是，由于收起遮盖物的过程非常麻烦，并没有将其付诸实践。

以前，要是建造一座新建筑，会将其很大一部分进行切割，填充到地下，避免展现在观众视野中。不过，在Termalija家庭健康中心项目中，建筑师不能简单地重复这种设计方法。必要的场所，特别是在有水的区域，需要一个更大的结构空间，不能将其简单地埋在地下，不能仅仅是参考周围的自然景观。恰当的解决方案应该是在周围乡土建筑的规模和形式中寻找灵感。

但是建筑师们仍然可以像设计他们的前一个项目Orhidelia一样设计Termalija的支撑结构，将水域上方的大屋顶分成几个部分，以防止规模超过周围建筑。从远处看，这个新的建筑群落结构的形状、颜色和规模都与周围乡村建筑浑然一体，这些乡村建筑在视觉上一直延伸到了建筑的中心。

本项目中对于建筑结构的巧妙设计为设计者解决了另一个难题。看似复杂的几何形状使得整个泳池空间视觉上看不到支撑结构，但实则是屋顶在提供强劲支撑。通过这种方式，新屋顶悬浮在泳池平台的上方，而泳池平台被设计成一个室外空间。由于夏季缺少露天空间，就不可避免地需要研究一些解决方案，以便在冬季能为游客提供封闭空间，而在夏季提供开放空间。屋顶上设有许多天窗，而且立面玻璃可以完全打开，这使得客人可以在建筑内部畅通无阻地穿行。尽管体积和占地空间大，但新的屋顶结构仅仅在泳池上方作为一个大型的夏季"遮阳伞"而存在，并没有侵占任何宝贵的室外空间。

Termalija Family Wellness is the latest in the series of projects which ENOTA has built at Terme Olimia in the last fifteen years, and concludes the transformation of the complex from a classic health center built in the 1980s to a modern, relaxing, thermal spa.

The new facility is located at the site of the former winter-time covering of the outdoor pool. The cover was supposed to be retracted every summer to free up the scarce outdoor space, but in practice this was never done because of the complexity of the retraction process. The existing winter covering was a detriment to the guests' experience of this otherwise very naturally designed complex.

Previously, a greater portion of all new buildings was cut-and-filled under the surrounding terrain with each subsequent project, which reduced their presence on the site.

In the case of the Termalija Family Wellness, however, the architects were unable to simply repeat this design approach. The volume of the necessary space, especially in the area with

项目名称：Termalija Family Wellness / 地点：Podčetrtek, Slovenia / 事务所：ENOTA / 项目团队：Dean Lah, Milan Tomac, Peter Sovinc, Nuša Završnik Šilec, Polona Ruparčič, Peter Karba, Carlos Cuenca Solana, Jurij Ličen, Tjaž Bauer, Sara Mežik, Eva Tomac, Jakob Kajzer, Maja Majerič, Goran Djokič / 结构工程公司：Ivan Ramšak s.p.
机电工程师：Nom biro / 泳池技术：Darrtech / 顾问：Intra Lighting / 主要承包商：Adriaing, Tipo, Lesnina / 客户：Terme Olimia / 用地面积：10,200m² / 建筑面积：3,930m²
总楼面面积：8,780m² / 造价：EUR 10,000,000 / 设计时间：2016 / 竣工时间：2018 / 摄影师：©Miran Kambič (courtesy of the architect)

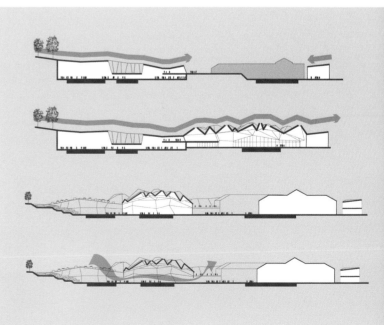

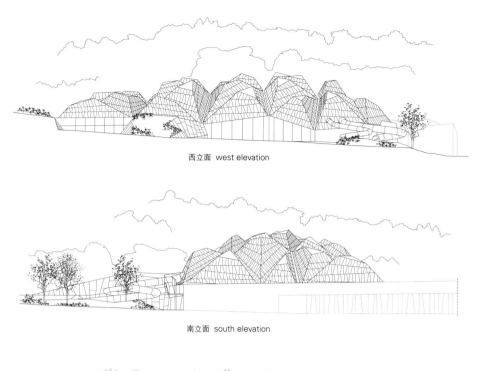

西立面 west elevation

南立面 south elevation

北立面 north elevation

一层 ground floor

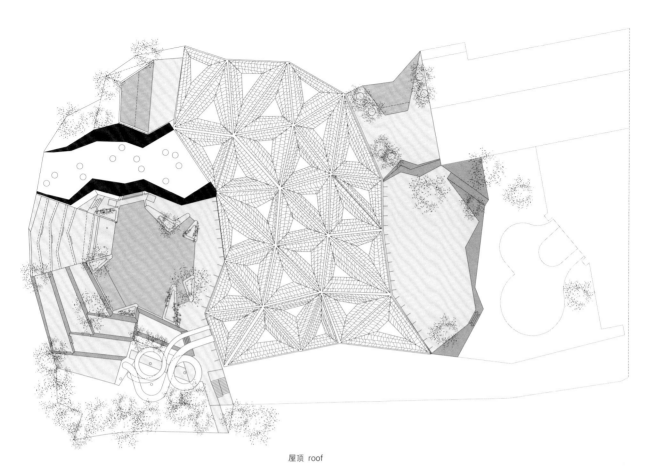

屋顶 roof

1. 休息场所 2. 游戏室 3. 儿童游乐室 4. 电影院 5. 保洁室 6. 护理室、小厨房、育婴室 7. 淋浴间 8. 卫生间、衣柜 9. 设备空间 10. 发动机房 11. 游泳池 12. 水上游乐场 13. 水上酒吧 14. 健身房 15. 室外露台 16. 露台 17. 家庭休息场所 18. 婴儿海滩 19. 护理室、卫生间、小厨房 20. 桑拿房 21. 滑梯入口
1. resting place 2. game room 3. children's playroom 4. cinema 5. cleaners storage 6. nursing room, kitchenette, babycare 7. showers 8. toilets, wardrobe 9. service 10. engine room 11. pools 12. water playground 13. aqua bar 14. fitness 15. outdoor terraces 16. terrace 17. family resting place 18. baby beach 19. nursing room, toilets, kitchenette 20. sauna 21. entrance to slides

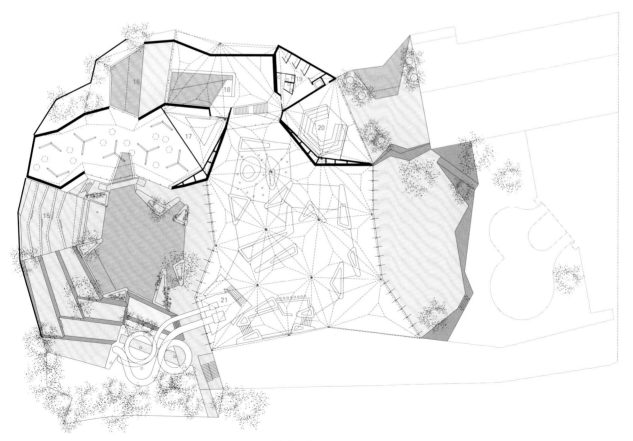

二层 first floor

water surfaces, required the siting of a much larger structure which could not simply be buried below the grade. No longer being able to reference only the surrounding natural landscape, the solution was found in the scale and form of the surrounding vernacular structures.

While the architects were still able to design Termalija's supporting program in the same way as they had done the previous project – Orhidelia – the large roof above the water area was divided into sets of smaller segments to prevent its scale from overwhelming the surroundings. Viewed from a distance, the shape, color, and scale of new clustered structure of tetrahedral volumes is a continuation of the cluster of the surrounding rural buildings, which simply extends, visually, into the heart of the complex.

The resulting fine structure of volumes has enabled the designers to solve another important task. The seemingly complex geometry gives the new roof static strength, allowing the entire pool space to be covered virtually without supports. In this way, the new roof floats above the pool platform, which is designed as an exterior space. Due to the lack of outdoor spaces in the summer, it was imperative to examine solutions which would deliver the same closed space used by the guests in winter as an open space in summer. Numerous skylights in the roof, together with the ability to fully open all the facade glass surfaces, enable the guests to seamlessly pass through the interior of the building. Despite its size and the space it occupies, the new roof simply acts as a big summer-time sunshade and does not usurp any of the precious exterior space.

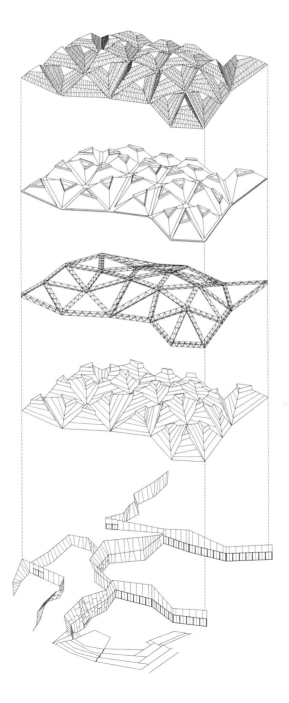

A-A' 剖面图 section A-A'

1. 游泳池
2. 水上游乐场
3. 儿童游乐室
4. 水上酒吧
5. 婴儿海滩

1. pools
2. water playground
3. children's playroom
4. aqua bar
5. baby beach

B-B' 剖面图 section B-B'

Pat Hanson

P198 gh3* architecture

Partners Pat Hanson and Raymond Chow lead an office team of twenty that includes architects, landscape architects, designers, and technical staff. They have been awarded Governor General's Medals for the Borden Park Pavilion in Edmonton and the Boathouse Studio on Stony Lake in Ontario. Pat has lectured on the work of gh3* architecture in Europe and North America, and has taught at University of Toronto and University of Waterloo. Was recognized by the international arcVision Prize for Women and Architecture in 2016.

P66 AZstudio

Guido Zuliani is an architect and an educator. Has founded Azstudio in 2003, and as principal he has collaborated in partnership with various firms in the United States and Italy. Graduated from the IUAV University of Venice and in 1985 began his teaching at the Irwin S. Chanin School of Architecture at the Cooper Union where currently has the position of Distinguish Professor in Architecture. Is also Affiliated Faculty at the Doctoral Program in Architecture, City and Design of the School of the IUAV University of Venice. Currently is Visiting Professor at the Department of Architecture, Built Environment and Construction Engineering of Polytechnic University of Milan.

P98 KAAN Architecten

Was founded at Rotterdam by Kees Kaan[left], Vincent Panhuysen[center] and Dikkie Scipio[right] in 2014. Expanded its international presence with satellite offices in São Paulo (2015) and Paris (2019). Kees Kaan was the founding partner of Claus en Kaan Architecten. Graduated at the Faculty of Architecture of TU Delft in 1987. Has been professor of Architectural Design at TU Delft since 2006. Vincent Panhuysen joined Claus en Kaan Architecten in 1997 and became partner in 2002. Studied at the Utrecht School of the Arts and the Rotterdam Academy of Architecture. Dikkie Scipio joined Claus en Kaan Architecten in 1995 and became partner in 2002. Studied at the Royal Academy of Art, The Hague and the Rotterdam Academy of Architecture. Was recently appointed Professor of Architectural Design at Münster University of Applied Sciences.

P66 Degli Esposti Architetti

Is founded in 2006 by Lorenzo Degli Esposti and Paolo Lazza, based in Milan. Their project is on the palazzo in urban and metropolitan contexts, thanks to both a broad practical experience and fundamental studies on historical precedents and contemporary examples, in Milan and other world cities. In 2015, Lorenzo Degli Esposti was appointed curator of the Architecture Pavilion for the 2015 Milan Expo.

Peter Eisenman

P66 Eisenman Architects

Peter Eisenman is an internationally recognized architect and educator whose award-winning large-scale housing and urban design projects, innovative facilities for educational institutions, and series of inventive private houses attest to a career of excellence in design. Currently the Charles Gwathmey Professor in Practice at the Yale School of Architecture, his academic career also includes teaching at Cambridge, Princeton, Harvard, and Ohio State universities. Holds a Bachelor of Architecture degree from Cornell University, a Master of Science in Architecture degree from Columbia University, and M.A. and Ph.D. degrees from Cambridge University.

P8 Peter Haimerl . Architektur

Peter Haimerl, whose architecture office was established in Munich in 1991, concentrates on projects, which cross the line of conventional architecture and demands fascinating and unconventional solutions to create innovation in every project. Currently dedicates his work to the leitmotif "attraction instead of restriction" and is interested in construction within existing structures. Is a professor at University of Art and Design Linz. Won a wide range of awards, including DAM Award for Architecture in Germany, culture award of the Bayerische Landesstiftung, Große Nike BDA.

P150 MUMA

The principals (McInnes[left], Usher[right], McKnight[center] Architects) met when they studied together at the Mackintosh School of Architecture in Glasgow and established MUMA in 2000. Their current public projects include a Masterplan for the Fitzwilliam Museum in Cambridge, the redevelopment of Abbot Hall Art Gallery and Museum in the Lake District and a gallery for modern art at Grimsthorpe Castle. Won Civic Trust Awards, including the Special Award for Sustainability, RIBA East's Regional, Sustainability and Building of the Year awards for their recent project Storey's Field Center and Eddington Nursery.

Peter Haimerl

P164 Querkraft

Was created in 1998 in Vienna. Have achieved about a hundred projects, competitions and constructions both in Austria and abroad and their team is composed by over forty collaborators. Their motto, "giving people space" is a way to express their will for spatial generosity. This desire has been particularly developed through the agency's housing projects. The U31 building for instance has received the Austrian state prize for architecture and stainability in 2013. They have also achieved various administrative and cultural buildings.

P190 Diego Terna

Graduated from Polytechnic University of Milan in 2004 and worked in the offices of Stefano Boeri and Italo Rota. In 2012 he founded the Quinzii Terna Architecture practice focusing on architecture, urbanism and research, supported by publishing, criticism and didactics. Contributes to national and international architecture magazines and websites. Taught in several universities in Italy, Central and South-America and China.

P44 OMA / Ellen van Loon

Ellen van Loon joined OMA in 1998 and has led several award-winning building projects that combine sophisticated design with precise execution. Some of her most significant contributions include Brighton College, BLOX, the new home of the Danish Architecture Center in Copenhagen, Rijnstraat 8, Lab City CentraleSupélec, and the G-Star Headquarters in Amsterdam. She is currently working on the Palais de Justice de Lille in France.

P4 Fabrizio Leoni

Is founder and principal of Fabrizio Leoni Architettura, based in Barcelona, started his independent practice while receiving his Master Degrees and PhD from SCI-Arc, Los Angeles and Polytechnic University of Milan, also attending Paris-Villemin, Architectural Association London, ETSA Barcelona. Currently teaching architecture and urban design at Polytechnic University of Milan, he has been lecturing and directing Design Studios at prestigious international institutions as Stanford University (California), Northwestern Switzerland University of Applied Arts, Tecnologico de Monterrey (Mexico), UI Jakarta (Indonesia), among others.

P178 Tezuka Architects

Was co-founded by Takaharu Tezuka[left] and Yui Tezuka[right] in Tokyo, Japan in 1994. Takaharu Tezuka (Tokyo, 1964) received his B. Arch. from Musashi Institute of Technology in 1987 and M. Arch. from University of Pennsylvania in 1990. Has worked for 4 years at Richard Rogers Partnership after graduation. Has been a Professor of Tokyo City University since 2009. Yui Tezuka (Kanagawa, 1969) received her B. Arch. from Musashi Institute of Technology in 1992 and continued her study at the Bartlett School of Architecture. Was a visiting faculty member at Toyo University and Tokai University. They gave lectures as Visiting Professors at the Salzburg Summer Academy and the University of California, Berkeley.

©Kristian Ridder Nielsen

P132 Paula Melâneo

Is a Lisbon-based architect. Graduated from the Lisbon Technical University in 1999, she received a MSc. in Multimedia-Hypermedia from the Paris Fine-Arts School in 2003. Besides the architecture practice, she also focuses her professional activity in the editorial field, writing critics and articles specialized in architecture for different international magazines. In Portugal, she was the editor-in-chief of *Jornal Arquitectos* (2015-2019); part of the editorial board of the Portuguese magazine *arqa Architecture and Art* since 2001 and its editorial coordinator between 2010 and 2016; editor for the *Architecture and Design Biennale EXD for 4 editions* (2011-2017) and the coordinator of several publications of the ExperimentaDesign association.

©Pierre-Olivier Deschamps / Agence VU

P82 Mei architects and planners

Was founded by Dutch architect, Robert Winkel in 2003. After studying construction economics at the University of Amsterdam, he graduated from the Faculty of Architecture, Delft University of Technology. Has worked for CEPEZED Architecten as project architect before founding Robert Winkel Architecten in 1996. Has been teaching post-graduate course at Delft University of Technology, Faculty of Civil Engineering, Department of Fire since 2004. Has been Graduation mentor at Delft University of Technology, Eindhoven University of Technology, Academies of Architecture in Rotterdam, Tilburg, Maastricht and Arnhem since 2000.

P210 Ferrier Marchetti Studio

Jacques Ferrier[left] and Pauline Marchetti[right] started collaborating in 2008, when they designed the French Pavilion for the Shanghai World Expo. Jacques Ferrier, following his architectural training at the École Nationale Supérieure d'Architecture de Paris-Belleville and the École Centrale de Paris, created his own architecture firm in Paris in 1993. Pauline Marchetti is graduated from the École Nationale Supérieure d'Architecture de Paris-Belleville, and a Professor at the École Nationale Supérieure des Arts Décoratifs de Paris. Their lecture widely at prestigious venues, including Columbia University GSAPP, Harvard University Graduate School Of Design and the AIA Center for Architecture.

Robert Winkel

Eran Chen

P118 ODA

Eran Chen is the founding principal of ODA. Since establishing the office in 2007 he has become one of the most prolific architects in New York. Graduated with honors from the Bezalel Academy of Art and Design in Jerusalem, where he serves on the board and as guest lecturer. Is a frequent speaker at design and development forums, as well as architecture schools including Columbia University, Clemson University, Syracuse University, Carnegie Mellon School of Architecture, and The Technion - Israel Institute of Technology.

P138 Malte Kloes Architekten

Malte Kloes[left] studied Design and Architecture in Potsdam, Berlin and Zurich. Received his MSc in Architecture from ETH Zurich in 2012. Has worked for several architectural practices including HCP Ltd. in Ahmedabad and BIG in Copenhagen. Collaborated with Christoph Estrada Reichen[right] on a variety of architectural projects between 2013 and 2018 and is running an independent practice in Zurich since 2019.

P26 Roldán + Berengué, arquitectes

Is an architecture studio in Barcelona, Spain, founded in 1988 by Miguel Roldán[right] and Mercè Berengué[left]. Is a multidisciplinary team of architects, engineers, artists, writers, and archaeologists from all over the world. Are experts in discovering promises in a problematic site, in a confusing program, in a low budget, in an incomplete product. For them all those problems are opportunities. A good problem is good resource. Specializes in sustainable residential architecture, corporate headquarters, public and educational, cultural facilities, rehabilitation of historic and industrial buildings, urbanism and planning.

P138 Estrada Reichen Architekten

Christoph Estrada Reichen grew up in Mexico City and Basel, Switzerland. Holds a BSc and MSc degree in Architecture from ETH Zurich and has worked in Switzerland and in New York City. In 2013, together with Malte Kloes, he founded a firm for architectural visualizations called "bildbau". In 2019 Christoph proceeded to incorporate his firm Estrada Reichen Architects.

P222 **ENOTA**

Was founded in 1998 in Ljubljana, Slovenia and is led by founding partners and principal architects Dean Lah and Milan Tomac. They graduated from the Faculty of Architecture, University of Ljubljana in 1998. Dean Lah was born in 1971 in Maribor, Slovenia. Has been a member of executive board of Chamber of Architecture and Spatial Planning of Slovenia, member of executive board of Architects Association of Ljubljana. Milan Tomac was born in 1970 in Koper, Slovenia. From 1998 to 2001, he was Assistant Professor at his alma mater. Received The International Architecture Award, Architizer A+ Award, IOC/IAKS Award and Europe 40 under 40 Award.

P164 **sam Architecture**

Was initiated in 2004 by a collaboration between Stefan Matthys and Boris Schneider. The agency was created in 2007 and has been led by Boris Schneider since 2009. Now concentrates mainly on public facilities and social housing projects. Over the years, the agency has gathered knowledge and experience about school architecture. Lead a theoretical and practical research on how education spaces are evolving, along with the teaching programs.

© 2021大连理工大学出版社

版权所有·侵权必究

图书在版编目(CIP)数据

守护城市密度 / 荷兰大都会建筑事务所等编；于风军等译. — 大连：大连理工大学出版社, 2021.5
 ISBN 978-7-5685-2975-4

Ⅰ. ①守… Ⅱ. ①荷… ②于… Ⅲ. ①城市建设—研究 Ⅳ. ①TU984

中国版本图书馆CIP数据核字(2021)第071021号

出版发行：大连理工大学出版社
　　　　　（地址：大连市软件园路80号　邮编：116023）
印　　刷：上海锦良印刷厂有限公司
幅面尺寸：225mm×300mm
印　　张：15.25
出版时间：2021年5月第1版
印刷时间：2021年5月第1次印刷
统　　筹：房　磊
责任编辑：杨　丹
封面设计：王志峰
责任校对：张昕焱
书　　号：978-7-5685-2975-4
定　　价：298.00元

发　行：0411-84708842
传　真：0411-84701466
E-mail：12282980@qq.com
URL：http://dutp.dlut.edu.cn

本书如有印装质量问题，请与我社发行部联系更换。